AF

The DJ Sales and Marketing Handbook

How to Make BIG Profits as a Disc Jockey

The DJ Sales and Marketing Handbook

How to Make BIG Profits as a Disc Jockey

Stacy Zemon

AMSTERDAM • BOSTON • HEIDELBERG • LONDON
NEW YORK • OXFORD • PARIS • SAN DIEGO
SAN FRANCISCO • SINGAPORE • SYDNEY • TOKYO
Focal Press is an imprint of Elsevier

ELSEVIER

Focal
Press

Acquisitions Editor: Catharine Steers
Project Manager: Heather Furrow
Assistant Editor: Stephanie Barrett
Marketing Manager: Lucy Lomas-Walker
Cover Design: Gary Ragaglia

Focal Press is an imprint of Elsevier
30 Corporate Drive, Suite 400, Burlington, MA 01803, USA
Linacre House, Jordan Hill, Oxford OX2 8DP, UK

Library of Congress Cataloging-in-Publication Data
Application submitted

British Library Cataloguing-in-Publication Data
A catalogue record for this book is available from the British Library.

ISBN 13: 978-0-240-80782-9
ISBN 10: 0-240-80782-0

For information on all Focal Press publications
visit our website at www.books.elsevier.com

05 06 07 08 09 10   10 9 8 7 6 5 4 3 2 1

Printed in the United States of America

Working together to grow
libraries in developing countries

www.elsevier.com | www.bookaid.org | www.sabre.org

ELSEVIER   BOOK AID International   Sabre Foundation

This book is dedicated to all disc jockeys who work, play, and live this profession. Share your passion with the world and recognize that true happines comes not from riches or praise, but from doing what you love. Great ability develops and reveals itself increasingly with effort. You will know you are on the right path when a series of naturally occurring and meaningful coincidences begin to happen as if the Universe were conspiring to assist you.

STACY ZEMON

Contents

Acknowledgments

The DJ Sales and Marketing Handbook reflects a synergy of many minds. Countless ideas discussed here owe their existence to serendipitous conversations that have occurred along the journey—especially in the hallways and seminar rooms at DJ conventions. I have also learned a great deal from the school of "hard knocks" called experiences and from reading trade publications and online articles written by my peers.

I am fortunate, indeed, to owe acknowledgments to so many superb people and am delighted to have this opportunity to publicly thank them.

First, last, and always, I want to acknowledge and thank God, the source of life without whom neither this book nor I would have been written. Please continue to guide me in properly using the talents, intelligence, and vision you have blessed me with, to be of service to others.

Moving into the human realm, I also extend my gratitude to the Focal Press publishing board for "green-lighting" yet another book from this DJ/author. Thanks, also, to all their publishing pros: Joanne Tracy, Emma Baxter, Catherine Steers, Stephanie Barrett, Becky Golden-Harrell, Christine Degon, Tricia LaFauci, Lucy Lomas-Walker, Abbie Jackson, and Sheri Dean.

I am, in particular, profoundly indebted to my dear friend Joan Wurzel. Her personal support and unerring editorial assistance have kept my ship on course, navigating through the chapters of this book.

Sterling Valentine, thank you for writing an incredibly inspiring foreword to this book. I am deeply honored by your kind words.

Kudos and thanks also to American DJ's marketing guru, Brian Dowdle, for coming up with the clever subtitle for *The DJ Sales and Marketing Handbook*.

It is with great admiration and sincere appreciation that I would also like to express my gratitude to the "best of the best" in our industry who have shared their know-how and opinions for the benefit of the readers of this book: "Paradise" Mike Alexander, Luis Barros, Randy Bartlett, Paul Beardmore, Ryan Burger, Bob Deyoe, Brian Doyle, Scott Faver, Jeff Greene, Scott Kiley, Chuck Lehnhard, Ray Martinez, Peter Merry, John Murphy, Matt Peterson, J. R. Silva, Kevin Howard St. John, Randi Rae Treibitz, and Sid Vanderpool. Thank you also to the many folks who provided photographic and marketing materials for this project.

Cheers! to Justin Owen, my Australian DJ buddy based in London. Your technical editing of the manuscript, as well as the information you shared with me about appealing to a global market, truly enlightened this author.

My appreciation also extends to my family of origin and choice (past and present), valued business associates, and spiritual mentors: Shayna Appel, Luis Barros, Joyce Cohen, Steven Covey, Douglas A. Cox, Brad Davis, Gordon Feller, Gene Goodman, Gloria Goodman, Russell Goodman, Rosemary Hargreaves, Patricia Ireland, August (Gus) Jaccaci, Patricia Johnson, Landmark Education, Shirley MacLaine, Jamey Marchese, Ian Mayo-Smith, Dick McDonough, Bette Moore, Charlie Parker, Matt Peterson, Gene Roddenberry, Robert Roth, Krishna Sondhi, William Spady, Diane Sullivan, Neale Donald Walsch, Robert Zemon, and Tom Zemon. You light up my life!

I also wish to acknowledge these fine companies for whom I am an Artist Representative: American DJ lighting, PVDJ audio, Promo Only CDs and DVDs, and Cord-Lox cable management. Thank you, Scott Davies, Brian Dowdle, Adam Lawson, David Ellefson, Jim Robinson, Pete Werner, and Dave Deavenport, for the opportunity to represent and use your fantastic disc jockey equipment, music products, and accessories!

Last (but certainly not least), I wish to acknowledge you, dear reader, for the investment you have made in this book. I hope you flourish and prosper in every way imaginable as a result of reading it!

Foreword

Don't bother reading this foreword.

I'm sure you know that the purpose of a foreword is to get you to want to read the book. Somebody tells you how wonderful the author is, how great the book is, and why you should read it, so you can stand there in the store wondering whether or not you should buy it.

Look, I'll be honest with you. Since you now hold in your hands what may be the most important book ever published in the DJ industry, it would be a complete waste of time to read what I have to say about it. What you should be reading is what Stacy Zemon has to say. Every single word of it.

After all, unless you're brand new to being a DJ (or you've just been living under a rock), you already know that Stacy is one of the foremost experts on running a successful disc jockey business. She's been a DJ entertainer for more than two decades. She's continually invited to speak at seminars and conventions all over America. She's consulted with other DJ services from around the globe. She's even been a radio DJ. Most importantly, she has written the world's best-selling "how-to" book in the industry, hands-down, which has touched the lives of literally tens of thousands of people.

But since just about everybody who's anybody in the DJ game knows who Stacy Zemon is, you don't need me to tell you.

In fact, if I were you, I would skip this foreword and start reading this book right now. If you really want to keep reading what I have to say about it, I can't stop you. What I can do as a professional marketing consultant and former mobile DJ is to tell you two very important things:

- The first thing is that Stacy Zemon knows what she's talking about. I've consulted with organizations from the Fortune 500 and government-agency level to the United Way and countless entrepreneurs and professionals. I don't tell you that to impress you—I tell you so you can feel confident in the material presented here. Trust Stacy.

- The second thing is that the value this book will bring to your business is worth a hundred times what you will pay for it—if you apply the information inside it.

But what would make somebody who already wrote *the* book on DJ'ing feel compelled to write another one? Why would she ask some of the most prominent disc jockeys in the business to contribute their opinions and know-how to this project? What could possibly be so important that Stacy decided it was worth going through the whole thing all over again?

I'll tell you a secret. It's something that the highest-paid DJs already know. The secret is this . . .

YOU COULD BE THE GREATEST DJ IN YOUR AREA, BUT IF PEOPLE DON'T KNOW ABOUT YOU, THEY CAN'T HIRE YOU.

It's simple and obvious, but don't be fooled. Just because it's easy to understand does not mean it's easy to fix. If it was, then every disc jockey would be supersuccessful.

You may have already discovered that your success is not determined by talent alone. This comes as a shock to some who believe that just being a "great entertainer" will automatically trigger a flood of referrals and attract all the clients in their marketplace. Unfortunately, that's just not the way it works. Marketing and sales—not talent—are the keys to unlocking your unlimited earning potential as a DJ business owner.

Luckily, learning how to market and sell your services effectively is exactly what this long-awaited book is all about. With the proper information, I believe you can do anything you set your mind to, but is it really that important to study this stuff? Can't you just "wing it"?

Well, believe it or not, there are still some DJs who haven't read Stacy's previous book, *The Mobile DJ Handbook: How to Start and Run a Profitable Mobile Disc Jockey Service*. Maybe that's why we still hear about horror stories like the DJ who showed up in casual clothes to an elegant affair, or the DJ who didn't have any of the client's special songs, or the DJ who brought a boom box instead of a sound system . . . or, worst of all, the DJ who just never showed up.

Did you ever try to dance while the disc jockey tried to beat mix for the first time? Ouch! See what happens when you don't know what you're doing?

The same is true about marketing. What you don't know *can* hurt you.

I hope you're beginning to see just how important this book can be for you. The truth is that there is enough business in your geographic area to make you rich. People are lined up, waiting to give you their money, but every time someone less qualified than you gets hired for a job that should have been yours, who's fault is that?

Remember, your prospects can only choose from the DJs they know about, and if they don't know about you, it's not their fault—it's yours. If you are a disc jockey who provides quality services, then you have an obligation to the people in your area to market yourself effectively. In fact, you are doing your prospects a disservice by not letting them know about you, and that's causing you to leave a lot of money on the table.

This is what I taught in my seminar at the 2004 DJ Times International DJ Expo in Atlantic City, when I first met Stacy. At the end of my presentation, some people were kind enough to come up to me and thank me. One of them was Stacy. Frankly, I'm just glad she came up after and not before, because it would have been a lot more pressure if I knew she was there!

I was so glad to hear that she was writing this book and honored when she asked me to write this foreword for her. The previous forewords to her first book and revised edition were written by Jim Tremayne, editor of *DJ Times* magazine, and Jim Robinson, director of Promo Only. These two guys are serious players in the disc jockey industry. I know they wouldn't waste their time endorsing something unless they were sure you could benefit immensely from it, and neither would I.

That's why I hope you understand how important your work is, and how important it is to reach those who are looking for you.

As a DJ, you have a rare and unique opportunity to create a once-in-a-lifetime experience for someone that they'll never forget. When you realize that the average person planning a party faces the daunting task of being the general contractor on a project with multiple vendors and deadlines costing thousands of dollars, you begin to understand the pressure, frustration, and anxiety your prospect may be facing.

Don't forget that club and bar owners face difficulties too. They want to hire the best talent so that they can attract the most patrons, but they can only choose from those disc jockeys they know about.

That's why they need you. You are talented. You are here to share your gifts and make a difference in the world, but you can't do that if you can't be found.

Since this is a sales and marketing book, it's about making more money. But when you market yourself effectively, you will have the opportunity to attract more clients and make a greater contribution to people, so it's also about what you can give. In fact, these two concepts are intertwined. The more you give, the more you get. The key to activating this powerful combination is to start giving.

Stacy understands this, and once again, she has given every DJ in the world a priceless gift. Buy this book, read it, memorize it, and most importantly, *use it.* I urge you to accept this gift that Stacy is giving you and to apply this information to grow your business. When you begin to treat your marketing and sales skills with the same respect, enthusiasm, and artistry as your disc jockey entertainer skills, your rewards will be unimaginable.

The marketplace is calling you. There are people waiting for you to rock their party and give them a story they can tell for the rest of their lives. Stop reading this foreword, and start reading the rest of this book! Your adoring public awaits you . . .

Sterling Valentine,
Marketing Consultant
www.SterlingValentine.com

About the Author

With love in her heart and music in her veins, Stacy Zemon's career nearly translates into DJ evangelism as she spreads her consummate knowledge on a global scale. *Mobile Beat* magazine calls her "one of the most successful women in the DJ profession." Her girl-next-door good looks and captivating smile instill a refreshing sense of Yin in the Yang-heavy DJ Kingdom. Zemon's creative wizardry has spun a 25-year career into an impressive record of success.

This woman has "WOW" power. Since 1979, Zemon has been revving up party-goers using a recipe for packing dance floors that she cooks up fresh at every event where she performs. Her upbeat, polished, and tasteful emceeing skills are second to none in the industry.

Zemon has authored a thoroughly well-scripted story of achievement as a club, mobile, karaoke, and radio disc jockey; business owner; consultant; and speaker. She has worked for four mobile disc jockey companies, started three of her own, and consulted to DJs in the United States and Europe. Zemon has also been an announcer, music director, program director and operations manager in Top 40, Contemporary Hit Radio, Adult Contemporary, Middle of the Road, and Country music radio formats.

A Multitrack Mind

Zemon's interest in music and entertaining began when she was a teenager in Connecticut. At age 12 she became a professional magician and booked herself on TV. A year later Zemon created an in-house (literally) radio station using the initials of her name as call letters on the now-archaic equipment that was available in the early 1970s.

During the summers of 1977 and 1978, she worked as the assistant social director at Tamiment, a major resort in the Pocono Mountains of Pennsylvania. Zemon ran contests and games and taught participation dances to vacationers.

From there she headed south to Nashville. Zemon first started spinning records at a Holiday Inn near her college campus, as well as on the radio. After a few years, Zemon moved to Philadelphia, where she learned the arts of being a mobile and karaoke entertainer. Her live entertainment gigs financially supplemented Zemon's work as a mid-day announcer at a radio station in Trenton, New Jersey, and then as the afternoon drive announcer at another in Philly. As the program director there, Zemon regularly listened to music product sent by record labels.

She Wrote *the* Book on DJ'ing

Several years later after moving back to the Connecticut suburbs, marketing-savvy Zemon authored the world's best-selling book on DJ'ing, which has helped many thousands of mobile DJs to build their businesses and incomes.

The Mobile DJ Handbook: How to Start and Run a Profitable Mobile Disc Jockey Service was first published in 1997 by Focal Press. Great popularity and new technology dictated that Zemon author a thoroughly updated version, which was released in late 2002. A Spanish version called *Manual del DJ Móvil* was published in 2003.

According to *Mobile Beat* magazine, "*The Mobile DJ Handbook* is not only a great resource for aspiring disc jockeys who desire to be successful in this highly competitive business, but also for those experienced pros who want to keep ahead."

Leader of the Pack

In addition to being one of the industries top entertainers, Zemon is known for her business acumen. Following the success of *The Mobile DJ Handbook*, this entrepreneurial DJ originated a ground-breaking concept by partnering with a national broadcasting company to form a mobile disc jockey entertainment division at a chain of radio stations. Using their call letters and advertising capability, it took less than a year for this multisystem operation to become the dominant force in Connecticut and Massachusetts.

Zemon eventually sold her business interest to her partner, which was then bought out by another radio broadcasting company. She moved north to Northampton, Massachusetts, and after a few years of living and entertaining there as a mobile and radio jock, Zemon moved back to the Philadelphia area in 2005.

Endorser of DJ Products

In 2004, Zemon entered into an artist-endorsement agreement with American DJ lighting, PVDJ audio, and Promo Only CDs and DVDs. She makes personal appearances at DJ conventions, and uses and endorses their products in her performances.

As part of the artist-endorser agreements, PVDJ (www.pvdj.com) and American DJ (www.americandj.com) have sold *The Mobile DJ Handbook* through their websites.

She's Road-Tested

Zemon has routinely shared her vast industry knowledge with her peers by speaking on the topics of success, leadership, sales, and marketing at DJ Times International DJ Expo, the Mobile Beat DJ Show & Conference, and the Mid-America DJ Convention. She has also written articles for *DJ Times* magazine and has been a judge at their Disc Jockey of the Year competitions.

Zemon has had a lifelong interest in exploring the mind-body-spirit connection. She is also committed to ongoing professional and personal education and is a proud member of the *American Disc Jockey Association.*

In addition to her DJ-related work, Zemon is a New England Regional Emmy award–winning television producer. Beyond the console, she is a social activist, combining her musical talents with her sense of societal mission to produce special event fundraisers for a variety of nonprofit and charitable organizations.

Undoubtedly, the secret to Zemon's success lies with her timing, determination, imagination, and deep commitment to increasing the professionalism and knowledge-base of the mobile disc jockey industry. Perhaps that's why *DJ Times* magazine calls her, "a veteran of the industry, who has contributed much to its evolution."

Currently, she runs Stacy Zemon Entertainment, is in the process of expanding DJ Camp® on a national level with two partners, and is both a sought-after consultant and speaker for a variety of different audiences.

Stacy Zemon invites you to contact her to share how the information in this book has helped you. You may also inquire about her speaking or consultation services by sending an e-mail to djstacyz@aol.com or visiting her website at www.stacyzemon.com.

Introduction

Time Travel Theater™
A Journey through Space and Time with Stacy Zemon

Welcome to Time Travel Theater! You have arrived at a magical location in space and time, somewhere between Possibility Street, Creativity Avenue, Realization Terrace, and Prosperity Lane. Here you will create an Innertaining and Entertaining performance filled with Soulful Stirrings, Humorous Explorations, "Aha!" Awakenings, and Life-Altering Revelations. The curtain is going up and . . .

It's Show Time. The spotlight is on YOU as the headliner in a live-action, distinctly original, and mind-blowing production of *The Greatest Show on Earth* . . . YOUR LIFE! The "real" you steps into the limelight, where you are joined by your co-stars. Each of you plays a part as the cast, crew, and audience. Performing together, your "hidden" talents and star qualities are revealed. All the resources you need to hit your mark are right here—the dialog, music, props, lighting effects, games, costumes, and direction from your inner guru.

The Whole World Is Your Stage. And your inspirational muse (yours truly) is here to assist you make this the passionately satisfying, powerfully useful, spiritually enlightening, and financially abundant performance of a lifetime! But it's your show, and only you can transform it from a tragedy or comedy into a musical love story.

The Plot Is Revealed. As Act I begins, you walk to center stage, look up, and ask, "Mirror, mirror, on the ball. Who is the greatest DJ of them all?" The "plot" is revealed as you *see* the Truth. You are a unique and original work of art who was divinely created, as an imaginative being, capable of magnificent accomplishments. You were born with all the equipment needed to achieve them. There is no stopping you but for your inner critic.

You are fully response(able) for scripting the circumstances in your life and for working toward the fulfillment of your dreams. You can be, do, or have everything you can imagine—and *believe* you deserve.

The Dramatic Conclusion. Life is a Party, and the future is NOW. You are the Music, so, let it flow from your heart and share it with all who will listen. As a DJ entertainer, you have an extraordinary opportunity to motivate and energize thousands of people. You have been gifted with the power to positively affect the audiences you entertain through your every word and action, as indelible impressions in their hearts and minds, both now, and for years to come. They freely give you thunderous applause to thank you for the joy, happiness, and fun you have brought into their lives. Do you know that you are surrounded by adoring fans who have come to cheer you on? Go ahead. Take a bow. Enjoy your standing ovation!

Looking forward to meeting you somewhere in space and time, I am . . .

Yours faithfully,

Stacy Zemon

A Note to Readers outside the United States

Dear DJ friend,

This book is intended to benefit an international community of English-reading DJs; however, the text is admittedly very U.S.-centric in its content. It is therefore incumbent on the reader to "translate" information that is not relevant or accurate within your locality. What's more, American spelling is used here (for example, words ending with a "z" rather than an "s").

My sincere apologies to you for having to put in this effort. The United States is where I have learned the art of DJ'ing. As my home ground, this is where my experience has been gained, and where I have obtained the vast majority of resources and information necessary to write this book.

I know the European Union is expanding and hope there will be another opportunity for me to take a more global approach—through either translations of *The DJ Sales and Marketing Handbook* or revised editions. Please know that I am sensitive to the content and language issues and would relish the chance to further address them.

Until then, please do go ahead and buy this book. I am extremely confident that you will find the majority of content here to be highly pertinent for you, regardless of where you live or what type of DJ you are.

Thanks for caring enough about your career to want to achieve success, grow your business, and get paid to party . . .

Cheers!

Stacy Zemon

*Formulate and stamp
indelibly on your mind a
mental picture of yourself
as succeeding. Hold this
picture tenaciously. Never
permit it to fade. Your
mind will seek to develop
the picture. Do not build
up obstacles in your
imagination.*

NORMAN VINCENT PEALE

Prosperity: Creating Inner and Outer Abundance

Creating inner and outer abundance requires placing your attention on more than just financial riches. It also includes having a passion for your career, engaging in productive business partnerships, giving back to your community, maintaining enriching personal relationships, and having faith (or some belief in a higher power). If you follow this daily regimen for prosperity, then the cumulative effect will be a lifetime filled with tangible and experiential riches both for yourself and those whose lives you have touched.

LUIS BARROS

Photo courtesy Luis Barros, LB Ventures, LLC

Managing Director: LB Ventures, LLC
Credits: Partner in DJ Camp, former DJ, formerly affiliated with LaFace Records and DARP/Cyptron, AFTRA/SAG member, board of directors of several nonprofit and private groups. Extensive entrepreneurial, investment, regulatory, and academic background in venture development and capital markets.
Academic: M.B.A., MIT-Sloan; B.B.A., University of Massachusetts at Amherst; Exec. Ed., Harvard University, Cambridge

Success Defined

The best disc jockeys constantly strive to improve their performance. They take pride in their work and are committed to their clients' satisfaction. They endeavor to achieve a level of excellence better than the best; they understand that success is based not only on talent, but also on persistence, drive, and business acumen.

Working long hours on your feet and maintaining a high-energy level is not easy. Achieving and maintaining a positive attitude can be a challenge. But each time you put on an outstanding performance, it will be noticed by people who can spread the word about the great job you did.

Success can be viewed in a variety of ways. Our society tends to define it based on outer "trappings" such as how much money a person makes, the size of his or her house, the type of car driven, etc. Another way to view success focuses more on one's inner experience of life. It has to do with how much joy, love, and happiness you feel. Success can even be viewed in terms of distance traveled—especially for people who have had many obstacles to overcome in their lives.

I believe the most useful definition of success centers on the continued expansion of happiness and the progressive realization of worthy goals. Successful people understand that there is no such thing as failure, only mistakes from which we can learn. They are persistent, dedicated risk-takers who have an unwavering belief in themselves and what they are doing. They aim high, learn quickly from their experiences, are action oriented, and love challenges.

Success is largely determined by our hard work and our choices. To be successful, you must act as if you are. At first, acting like a successful person may feel as though you are an actor playing a part. Eventually, the "acting" will translate into "being." Demonstrate an attitude of success in every area of your life, such as your manner of dress, the way in which you speak with people, and how you conduct yourself and your affairs.

Highly successful people not only focus on a "destination" or goal, but they also enjoy the journey involved with getting there. Before you set out on your journey, you ought to know something about where you want to go and when you expect to arrive.

One of the most important predictors of your success is attitude. Self-confidence is achieved by knowing who you are, what you have to offer, and by eliminating the factors that cause self-doubt. Be enthusiastic! Verbalize your excitement about your business to everyone with whom you come in contact.

Successful people have the ability to focus on the positive and have faith that their goals will manifest themselves. Faith is very different from hope. Faith is the inner knowing of things unseen and a steadfast belief that there is a power greater than yourself who can assist you. Your goals can be actualized through the use of prayer, meditation, affirmations, and action.

Energize your mind with books, DVDs, music, and activities that will motivate, relax, and inspire you. Eating healthfully and getting regular exercise will also help keep you in peak physical condition.

Maintaining a heart filled with gratitude is an excellent way to open yourself to receiving. Gratitude is the Law of Increase. Expressing gratitude can be as simple as a sincere "thank you" to God and to the people with whom you come in contact on a daily basis. Be grateful for everything that you receive.

Inner and Outer Abundance

Perhaps you are holding onto an idea of what financial wealth should be for you, based on some vague or even specific idea. How much money would it take for you to consider yourself "wealthy?" $50,000? A million dollars? More?

Actual wealth is not limited to money. Abundance can come in ways we do not recognize and therefore fail to acknowledge. Many people "wish" they were wealthy, without realizing the wealth already present in their lives. True affluence also manifests as an abundance of good relationships, health, talent, and creative inspiration as well as qualities of character such as compassion, wisdom, and generosity. By letting go of fixed ideas about wealth, you empower yourself to start enjoying the consciousness of abundance, an inner state that draws wealth to it.

An important aspect of building an experience of wealth in your life is to put some of your income aside in a savings account. It may seem strange when you are struggling to pay your bills, but even if you owe more than you can pay in any given month, knowing that you are building a "nest egg" gives you a real feeling of abundance. It's kind of like paying yourself first before anyone else to whom you owe money.

 MY COVENANTS FOR SUCCESS

**I pledge daily to read this list out loud,
and that to the best of my ability I will:**

1. Create and focus on my S.M.A.R.T. (Specific, Measurable, Achievable, Realistic and Timely) goals, and believe that I can achieve and deserve them.

2. Seek through prayer and meditation to connect with God or the Higher Power of my understanding.

3. Ask God for help in achieving my goals, and in letting go of any fear or resentments I may harbor.

4. Pursue and secure at least one mentor, one coach, and one spiritual advisor to support me in achieving my goals.

5. Faithfully work with them and others to continuously develop my talent, knowledge, skills, and abundance.

6. Operate with honor and integrity, doing what I love and loving what I do.

7. Conduct a searching and fearless personal inventory, and admit to myself, God, and my spiritual advisor the exact nature of any actions I have taken in my life for which I feel guilty or ashamed.

8. Make direct amends to any persons I have harmed if possible, except when in doing so I would knowingly cause harm to them or others.

9. Continue to take a personal inventory and when I am mistaken, I will promptly admit it.

10. Express generosity and gratitude through my thoughts, words and actions to everyone with whom I come in contact.

11. Develop a net worth of $_____ by _____, wisely saving _____% of my monthly income, and utilizing every legitimate tax management advantage available.

12. Act as if I have already achieved great success as the result of these steps, ever practicing these principles in all my affairs, and sharing my wisdom with others.

_____   _____
Signature   Date

Written by Stacy Zemon, July 16, 2005

If you do have debts, be sure to pay back some amount on a regular basis, preferably monthly, no matter how small it may seem. You are building a flow in consciousness that demonstrates to yourself and others that you can manage material wealth. As you begin to feel wealthier, you will actually attract more abundance.

Be charitable in your thoughts, words, and actions. As in so many areas of life, by helping to lift up others we lift ourselves up. Build up the circulating flow of generosity. After all, "what goes around, comes around." The goodwill and generosity that you put forth will come back to you many times over. It is the same principle in nature as planting: From one seed taking root and growing into a tree, countless seeds come forth.

Goal Setting

Baseball legend Yogi Berra once said, "If you don't know where you're going, you'll probably end up someplace else." A goal is a way to make sure you don't wind up someplace other than where you want to be. With your business, you must determine the people, financing, advertising, marketing, and other resources required to achieve your goals.

There are many ways to define a *goal,* and numerous synonyms for the word—*target, purpose, objective, destination, intent,* and *aim,* to name a few. Goals give us direction and focus. They turn impossible undertakings into achievable tasks. They help us keep our vision clear and our footing steady.

The process of setting goals forces you to think through what you want from your DJ service. Goals also give you a framework within which to work. A very important part of that framework is a timetable, which will influence your actions profoundly. For instance, if your goal is to earn $100,000 annually by age 30, then your growth plan must strategically plot precisely how you will accomplish this goal.

When setting your goals, make sure they have the following qualities:

- *Specificity:* Earning money isn't a goal. Earning $100,000 over the next twelve months is a goal.
- *Optimism:* "Not being homeless" is not exactly an inspirational goal. "Achieving financial security" phrases your goal

in a more positive manner, thus firing up your energy to attain it.

- *Realism:* If you set a goal to be a part-time DJ *and* a part-time astronaut, that goal is unrealistic. If you really want to be an astronaut, start by calling Space Camp and asking them to send you some literature.
- *Short- and Long-Term Goals:* Short-term goals are attainable in a period of weeks to a year. Long-term goals can be achieved in one, five, ten, or even twenty years from the present. They should be substantially greater than short-term goals.
- *Lifestyle:* This includes areas such as travel, hours of work, and investment of personal assets. How many hours are you willing to work? Which assets are you willing to risk?
- *Honesty:* Build a business with your eyes wide open about your strengths and weaknesses, your likes and dislikes, and your ultimate goals.

Keep a written record of what you are trying to accomplish. Communicate those goals with significant others in your personal and business life, as well as with a success coach. Brainstorm with knowledgeable people to find creative ways to meet your goals.

Creating an Action Plan

A necessary ingredient in any formula for success is vision—your underlying, driving, desire-filled concept of what you value most in life. To convert a vision into a goal, break it down into a workable action plan of daily, weekly, and monthly steps.

First, write out a vision statement. Detail for yourself what you want from your business and what will motivate you to take action. You can be as specific as you like and include questions relevant to all your hopes and dreams. If the sky were the limit, what would you want? Write a list of your emotional, intellectual, physical, spiritual, and monetary goals. Set short-term and long-term goals. Review these goals daily, and visualize yourself achieving them.

When your list is complete, eliminate all the items you wrote down because you thought you should, not because you really want or need them. Ask yourself if you are willing to make a plan to attain or achieve each of the remaining items on your list.

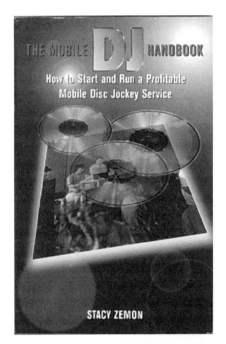
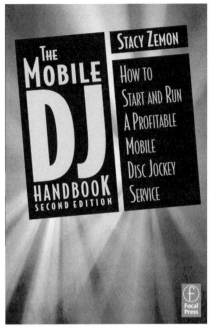

The Mobile DJ Handbook published in 1997. The 2002 revised second edition. The 2003 *Manual del DJ Móvil* Spanish translation.

If you love being a mobile DJ and focus on service, financial rewards will naturally follow. The focus must be on the service, not on the money. Financial rewards are always the secondary outcome of serving others well.

If you believe that someone or something outside of yourself is preventing you from succeeding, you're giving away your power to that someone or something. You're saying, "You have more control over my life than I do!"

Most certainly there are aspects of life you can't control, such as other people or "Mother Nature." What you can control, however, are your own thoughts and your actions. Taking full responsibility for your life is one of the most empowering things you can do for yourself.

Be willing to pay the price for your dreams. Wanting a big house, a luxury car, or a million dollars in the bank is fine, but the really important question is "Are you willing to pay the price to get them?"

Every day, we are bombarded with hundreds of tasks, messages, and people all competing for our time. This is why the ability to focus on your goal is so critical to achieving it. Focusing requires giving up some things in the present because you know the time invested will pay off big-time down the road.

The first requirement for business success is the habit of planning. The more thoroughly and detailed your activities are planned in advance, the faster and easier it will be for you to carry out your plans and obtain the results you desire.

Spend time each day focusing on achieving your goals and dreams. Ask yourself, "Is what I'm doing right now bringing me closer to my goal?" If it's not, do something that will. Focusing is like any habit; the more you do it, the easier it gets.

Taking the time to write out an action plan, or map, for how you're going to achieve your goals is one of the best ways to get there quickly. Goals that are not in writing are not goals at all. They are merely wishes or fantasies.

Second, create benchmarks or milestones of success so you can measure your progress. Remember that when you're in the process of learning, the going often seems slow. You need a way to measure your expanding knowledge and experience so you can affirm your own growth and development. Each day you may even want to write down "tiny victories." Recognizing, affirming, and celebrating your progress and successes will keep you motivated to meet the next milestone.

It may sound simple, even obvious, but when you're truly committed to achieving your goal, giving up isn't even an option. You must be willing to do whatever it takes to make them happen.

Your Teammates

Much of your success will be determined by the quality of the people who are your teammates. When it comes to your business, you may be used to doing everything yourself. As you expand your operation, the task will eventually become too large for just you alone.

At that point you must learn to think in terms of getting things done through others rather than trying to do it all yourself. This is the best way to leverage your time, and spend it in the areas that best serve the profitability of your business.

When you've delegated a task, set up a system of reporting so that you're always informed as to the status of the work. Be sure the other person knows what is to be done, and when, and to what standard. Your job is then to make sure he or she has the time and resources necessary to get the task done satisfactorily.

In addition, the people on your team need to know what's going on. The more thoroughly and accurately you report the details, the more support you will get and the happier they will be.

Two all-important members of your team are a mentor and a coach. Mentors can be living or deceased. They are people who exemplify business success. You can have more than one mentor. A success coach is a professional you hire to assist you in achieving your goals.

Overcoming Obstacles

Start thinking of obstacles as invaluable opportunities to grow, gain insight, and act creatively from a situation. I'm a firm believer that faith, in partnership with creative problem-solving and personal support, is really the only way to overcome obstacles.

When you catch yourself thinking negative thoughts—think again (literally). If you think you are in a bind or even a terrible financial situation that no good can possibly come from, think again. If you think your life is falling apart, and it looks as if you'll never get it back together again, think again.

The vast majority of the most successful DJ entertainers (including me), experienced their greatest successes after their biggest personal and/or professional setbacks. Through maintaining a positive attitude and sheer persistence in action, major accomplishments are not only possible—but inevitable.

Becoming a Master

How can we become masters of our lives? The process begins with awareness, noticing the patterns that continue to emerge in your experience time and time again. You may notice certain similarities in specific types of experiences. The names, faces, and exact circumstances may differ, but the underlying quality to the situation is very familiar. It's like having a déjà vu experience, a "knowing" that you have been in that experience before.

There are times in our lives when we all feel like we are "spinning our wheels" or "running around in circles." Perhaps these metaphors were created to describe the human condition of feeling "stuck" within an unproductive pattern of thinking and behaviors.

If a particularly unpleasant situation is continually repeating itself in your life, the only way to alter the pattern is through its recognition, acceptance, and a willingness to make the necessary changes. When a person is truly ready, this can happen in a moment of personal enlightenment when the "light bulb" comes on.

Many people have an ongoing fear of financial insecurity regardless of how much money they have. So, how do you overcome this fear?

- Understand that financial security is a feeling.
- Take an honest and detailed financial inventory.
- Make positive changes in your personal and professional life.
- Do prosperity affirmations (for example, I am financially secure, I am living my dream, I have plenty of time and money, and so on).

It is very easy to label experiences we do not particularly care for as "bad" or "negative." When we do this, we may miss the gift or lesson in the experience. Instead of judging the situation, try asking yourself:

- What did I gain physically, mentally, emotionally, and spiritually from the experience?
- What is the underlying message for me?
- What thinking or actions of mine contributed to creating the experience?
- What do I need to do to learn or heal from it?
- What changes can I make that will likely prevent this from happening again?

Searching for and reaching the answers to these questions are critical for developing an awareness of both self-destructive and beneficial patterns in our lives. It allows us to break long-standing personal behavioral cycles. If we no longer wish to repeat an experience, then we must become master of it. We may not always be able to change a situation outside ourselves, but we can choose to alter how we think and feel about it.

Here are some suggestions for mastering unbeneficial, repeat patterns in your life:

- Think deeply about how you participated in creating the circumstance.
- Have a sincere desire to break the pattern.
- Pray and meditate about it.
- Ask for help from God or the higher power of your understanding, and from supportive others to see the situation clearly.
- Make a conscious choice to think, feel, and behave differently than you have in the past, starting now.

Leadership

If you role-model leadership and set a good example within your DJ company, you will create a "corporate culture" that is positive for everyone who works for you. Being an effective leader will make everyone want to work hard to provide good performances. They will take pride in working for an organization that is a first-class operation with a distinguished reputation.

In their book, *Total Leaders,* Dr. William G. Spady and Charles J. Schwahn outline five domains of leadership. Professionals in any field, including DJs, can use this model as a self-assessment tool in order to become a "Total Leader."

Authentic Leaders

- Have a clear sense of personal and organizational purpose
- See themselves as "lead learners" rather than teachers
- Model core values such as honesty, responsibility, awareness, and gratitude

Visionary Leaders

- Define and pursue a preferred organizational future
- Maintain consistent client focus
- Expand perspectives and options for the organization

Quality Leaders

- Develop and empower everyone in the organization
- Consistently improve performance standards and results
- Create and use feedback loops to improve performance

Cultural Leaders

- Involve everyone in productive change
- Develop a change-friendly culture of innovation, healthy relationships, quality, and success
- Create meaning for everyone

Service Leaders

- Support and manage purpose and vision of the organization
- Restructure as needed to achieve results
- Recognize and reward positive contributions to productive change

Communication Skills

Leaders need to listen well to communicate effectively. Being a good listener and understanding a person's values begins with respect. Respect the person with whom you are communicating, of whatever age or background. Focus on hearing what the other person is saying, and be attentive to how he or she acts.

To enhance your relationship with your staff members and avoid misunderstandings, it is important to have good communication skills, including listening. To be a good listener:

- Pay attention and don't get distracted. When feasible, stop whatever you are doing in order to focus on the person speaking with you.
- Don't pretend you are listening when you are really not. If the timing is bad, tell the person it is not a good time to communicate at the moment. Make a specific follow-up time and be sure to follow through.
- Stay focused. We often think we are listening, when in fact we are rehearsing what we will be saying next. We worry about our remarks and practice them to make sure we get our point across. Other times we simply get lost in our own thoughts or actions.
- Remember that words have many meanings. People may take your words literally. Almost any message can be interpreted in several ways. We cannot assume just one meaning for the words heard.

Learning Styles

Everyone has greater aptitude in certain areas more than in others. It is helpful to know which aptitudes are your strongest so you can focus on activities in which you can excel. Being aware of the areas where you could benefit from improvement also can be helpful to your personal growth.

Dr. Howard Gardner, a professor of education at Harvard University, developed a theory of multiple intelligences. He proposes eight different intelligences to account for a broad range of human potential. These intelligences are as follows:

- Linguistic intelligence (being "word smart")
- Logical-mathematical intelligence ("number/reasoning smart")
- Spatial intelligence ("picture smart")
- Bodily or kinesthetic intelligence ("body smart")

Former American Disc Jockey Association (ADJA) President, Peter Merry, receiving the newly established Peter Merry Leadership Award at the 2005 Mobile Beat DJ Show and Conference in Las Vegas, Nevada. Photo courtesy www.ProDJ.com.

- Musical intelligence ("music smart")
- Interpersonal intelligence ("people smart")
- Intrapersonal intelligence ("self smart")
- Naturalist intelligence ("nature smart")

When our potential is actualized, and we are doing what we love and loving what we do, great success naturally follows.

Meditation

Some people think of meditation as listening to God and praying as talking to God. People achieve this by different routes, but all such methods have the effect of shutting down the thinking apparatus, which then allows deeper levels of consciousness to become aware.

Meditation is a practice that has been around for literally thousands of years. It is a progressive settling down of the mind into a field of silence and profound rest. Brainwave patterns change during meditation to a distinct pattern different from the normal awake state or sleeping state.

Meditation increases feelings of connection. It can also help most people feel less anxious. The awareness that meditation brings can be a source of personal insight and self-understanding.

Many people live in a state of perpetual motion and expectation that prevents them from appreciating the gifts that each moment gives us. Meditation makes us more stress resistant and reverses the physiological damage of stress. It affects our emotions, mental capacity, and chemical balance. It directly influences how our immune system works.

Meditation is a way to achieve comprehensive self-improvement in the realm of the mind and body, and at the same time it is a spiritual journey.

When you meditate, you might try visualizing the subject matter with yourself in the center of it. Consciously create this "movie in your mind" in the here and now, and use all your internal senses to make it as "real" as you possibly can imagine. You are the actors, director, producer, and set designer. As such, you can play out the scene anyway your heart desires.

I find that the harder I work, the more luck I seem to have.

THOMAS JEFFERSON

2

Business Matters: The Foundation of Your Castle

If you want your DJ service to stand apart from your competition, then strive for excellence, always pushing yourself to a higher level. Think about it this way, when you go the extra mile there is never a traffic jam! Make it your business to do the little 'extras' for your clients.

JOHN MURPHY (a.k.a. JOHNNY JAMES)

Photo courtesy John James Murphy, Eventscape America

Founder: Eventscape America, Star DJs, Star DJ's School of Live Entertainment
President and CEO: Inventures Corporation
National Spokesperson: Pioneer Laser Entertainment. Featured in a celebrity profile ad campaign for MTX Sound Craftsman mixers as "The Greatest DJ on the Planet." Led the national Diet Pepsi "Uh Huh" month promotion with Ray Charles and The Raylettes.
Media: Featured on CBS Morning News, and in the *New York Times* and *Newark Star Ledger* newspapers. Profiled on the American Entrepreneur series on FNN.
Producer: Let's Dance video series, "How to Be Successful in the Mobile DJ Industry" video and workbook program
Speaker: DJ Times' International DJ Expo
Organizer/Host: MTV's Spring Break "Rock a Like" competition
Recording: "I'm a Jammer" and a remake of "Shake, Rattle & Roll"
Recipient: CAPE award for TV programming excellence for "Star DJ's Party Club"
Celebrity Clients: Donald Trump, Katie Couric, Peter Jennings, Bruce Springsteen, Regis Philbin, George Strait, Kenny Chesney, and Lou Piniella

Industry Statistics

There is little verifiable information on exactly how many disc jockeys there are in the United States, let alone throughout the rest of the world. The estimated size of the U.S. market ranges between 50,000 on the low end to 200,000 on the high end. These figures include mobile, club, karaoke, turntablist, and remixer disc jockeys that perform in a wide variety of venues. The data are based on surveys conducted in the United Staes with DJ magazine publishers, DJ equipment manufacturers, DJ associations, DJ Internet site owners as well as music merchandisers, and affiliate associations.

According to some sources, the disc jockey industry has increased over 1,000 percent over the past ten years. Because of the growing interest in the DJ culture, an increasing number of formal disc jockey schools have sprung up around the world. In America, DJ Camp® offers programs for youngsters ages 10 through 17, who want to learn about the art of DJ'ing (www.djcamp.com).

According to a 2004 report from the National Association of Mobile Entertainers (NAME):

- 20% are full-time entertainers who treat their businesses as their sole source of income. Their company is a full-time business and usually represents and contracts multiple entertainers for multiple engagements each week.
- 60% are part-time entertainers who also treat their businesses as a full-time venture but who also usually have a second source of income. Some work as often as those in the first category.
- 20% are strictly part time. These entertainers most often work to supplement the income from their primary occupation and do not work as often as do DJs in the other categories.

According to the American Disc Jockey Association (ADJA):

- Disc jockey rates vary based on talent, experience, emcee ability, service, coordination, equipment needed, music knowledge, mixing ability, and personality.
- Rates for the DJ industry vary greatly, ranging from $350 to $3,500 for a four-hour booking.

- A full-service disc jockey company will normally invest 12 to 30 hours in planning an event although it may appear as if a client is only paying for "4 hours." Consultations, music purchasing and editing, preparation, setup and tear-down, education, and other business-related endeavors add up to the overall success of a special occasion.

According to www.ProDJ.com:

- The associated DJ equipment industry is estimated to be around $200 to 300 million worldwide, of which the US market is by far the largest segment.
- There is a very high correlation of purchase intentions with brand awareness and unaided recall. Perhaps the high correlation hinges on DJ sensitivity to magazine articles, reviews, and other forms of public relations.
- The main sources of information for the DJ market seem to be heavily concentrated on two publications: *Mobile Beat* magazine and *DJ Times* magazine.

Surveys taken by several wedding publications and websites show that:

- The entertainment chosen for a wedding or party is responsible for 80% of the event's success.
- 72% of today's brides are using professional mobile entertainers for their entertainment choice.
- 63% of the guests attending these functions do not remember what was served for dinner but do remember the entertainment.
- 81% of guests say the thing they remember most about a wedding is the entertainment.
- 4% to 6% is the amount most brides reserve for their entertainment budget.
- After their reception, 72% of all brides say they would have spent more time choosing their reception entertainment. Almost 100% say they would have spent more of their budget on the entertainment.
- During wedding planning, brides say their highest priority is their attire, followed by the reception site and caterer— reception entertainment is among the least of their priorities.

Within one week after their reception, 78% of brides say they would have made the entertainment their highest priority.
- 65% of all couples that chose a band to entertain at their wedding, said that if they had it to do over again, they would have chosen a disc jockey.

According to a 2004 article published in the *New York Times* newspaper,

> Until recently, aspiring DJs had to rely on a combination of osmosis and experimentation: You'd take mental notes at nightclubs, then you'd retreat to your room and keep practicing until you got the hang of it. Now, more and more people are learning how to DJ in classrooms.
>
> These classes alone don't make anyone a good DJ: even the most eager students know that mere competence requires months—not hours—of practice. But just as learning how to play the piano was once part of any serious listener's musical education, learning how to DJ, even if it's only an introductory lesson, may now serve a similar purpose.

Calculating a Marketing Budget

Your marketing budget should be based on your gross annual or projected gross annual revenue. Multisystem/multilocation operators generally need to spend more money on advertising than single-system operators because they have more jobs to fill.

Devote 5 percent to 25 percent of gross sales to your marketing budget, depending on your length of time in business. The longer you are in business, the less money you will need to spend in advertising, because your word-of-mouth referrals will grow.

Obviously, you cannot afford to spend more than you have. At 15 percent, a business projecting $50,000 in gross sales in a given year would spend $7,500, or about $625 per month, on marketing. As a rule of thumb, avoid spending more than you can easily carry as a monthly payment until the income starts to flow. Until then, you may need to supplement your income with a part-time or full-time job.

Your investment needs to be divided among your advertising, marketing, public relations, and customer service expenses. In your first year of business, these expenses will be greater because of the design and printing of your materials. Here is a suggested estimate to use for planning purposes:

- Year 1 in business—20% to 25%
- Year 2 in business—15% to 20%
- Year 3 in business—10% to 15%
- Year 4 in business—5% to 10%
- Year 5 in business and thereafter—5% to 10%

Remember, as your profit goes up, the percentage can go down, while the dollars remain the same or increase. So don't worry, you are not skimping on this important budgetary item.

Return on Investment (ROI)

You'll get the maximum return on investment by targeting your advertising messages at the people who are most likely to hire your disc jockey services. This is your target market. Whatever methods of advertising you choose, consider them a test. Develop a system for monitoring the results of each media (cost versus return).

Each time a potential new client calls, inquire how he or she heard about your company, and be sure to write down the response. (See the section titled "Tracking Your Results," later, for more detail.) Stay with what works best for your company, and if an ad or media is not providing sufficient ROI, make alterations accordingly.

The best measurement of an ad's effectiveness is whether it brings in enough new business to pay for itself at least twice over. It is important to judge media effectiveness not only by the number of inquiries you receive but also by the cost of obtaining those inquiries (for example, *Yellow Pages* versus local weekly newspaper, versus local cable print ad).

If five different DJ services advertise in a local newspaper that has a circulation of 120,000, those five disc jockeys are locked into tight competition. By placing an ad in a newspaper with a smaller circulation in which you are the only DJ, you will get all the leads from that paper. It is clear in this case which is the more cost-effective medium. Or you may choose to buy a more expensive package in a local cable show or on the radio, if it can generate a greater response rate. This is also a way to maximize your return on investment.

Generally, a run of four to eight editions of the medium gives you time to assess how well the ad is working. (For proper judgement, a monthly ad may have to run over a good part of the year.) If, after this time period, there has been little or no response, it is time to change the ad or the medium. One way to know which is to change the ad midway through the run to see the response. If the medium is working, keep the more effective ad running.

With targeted advertising directed to a few key media outlets, you will have an opportunity to bargain for discounts and stretch your budget. Promising longer terms for ads will also get you a bargain. If you contract to keep the ad in for a year, you will often pay less than four or five individual monthly ads. Be sure to ask your current representative for the ad rate sheet to check on this. To receive a discount (usually 10 percent), make it clear that you are booking for yourself and want the agency fee out of the pricing.

If your advertising is effective, it will lead to inquiries that you convert into clients through sales. In the event that the number of responses to your ads is greater than your ability to cover all the business generated, make sure to have a contingency plan in mind. Recommending another DJ service may seem counterproductive, but offering to do this with overflow business can lead to that service doing the same for you at slow times in your calendar. It is common for members of local DJ associations to have this type of arrangement. Just be sure to check out any service you recommend, so that potential future customers are pleased with your advice.

Purchasing Media

Most newspapers offer copywriting, typesetting, and graphic design for free or for a small additional charge to the cost of your ads. Avoid any outlet that asks you for large design fees. Radio and television stations, like print media, offer discounts for large media buys, and they will definitely steer you to a low-cost source for production. The amount of a discount depends on the total amount you spend.

Purchasing a "package" will provide you with significant savings over the cost of buying individual commercials. Packages usually include a combination of prime time and non–prime-time commercials, aired a given number of times per week for a specific number of weeks. You are not required to run them within consecutive weeks, as long as you use up the right number of commercials before the end date. Packages are sometimes bundled with an on-air promotion. The price you are quoted is generally negotiable by as much as 20 percent or more. Media account executives will try to offer you a larger number of commercials rather than a dollar discount. However, you will have additional leverage if you offer to prepay for your commercials. Remember, you are the buyer, so do not be afraid to ask for whatever type of consideration you need to make the media buy. Salespeople will do everything they can to make the initial deal, because they want you to be happy and to be a long-term customer.

Tracking Your Results

It is important to carefully track your advertising results, to effectively gauge how each medium is working, and to plan future expenditures. When tracking ads or commercials, you need to record how many prospects come to you through that medium, how many of them become clients, and how many become repeat clients. You also need to carefully record each dollar you spend on each ad type and the dollars brought in by it. Although this sounds tedious, it is the only way to make sure your ads are doing their job.

Ask every prospect where he or she heard of you, and take note of responses. Using a program such as Microsoft Excel, list all the options for hearing about your business and record the daily figures for each method. Beside each option, record the number of prospects, the number of sales, the number of exposures (how many times you ran the ad), and the cost of the advertising and its dollar return. Remember to include referrals and word-of-mouth in your list of options.

When prospects call, always ask how they heard about you and keep track of their responses. During conversations you can also gain valuable insight about potential customers that will give you an accurate picture of what is working and what is not. Also make note of how many inquiries you are converting into bookings. Here is the primary information you want to know:

- What ad prompted the person to contact you
- The geographic areas from which you are receiving responses
- The sex and approximate age of callers
- What type of event is being planned

At the end of each month, total up the daily figures and see where your best results are coming from. This information is the basis of future advertising decisions. When testing headlines and copy for ads, change only one thing at a time for realistic results. Also, use your account executive as a resource. He or she is usually quite willing to make suggestions and tell you what your competition is doing with its ads. Also ask the account executive's advice on proper placement of print ads on the page and, in the case of radio and television, on appropriate shows to insert your commercials.

Using Advertising Agents

Proper media selection and placement are areas that require professional expertise. Therefore, if you have budgeted more than $2,000 monthly for your advertising, consider retaining the services of an advertising consultant. When you consider the amount of aggravation you will save yourself, it is definitely worth a small percentage of your total ad budget. The time involved in buying and planning advertising can work out to be a full-time job in itself, so try to cover the cost of a small agency through increased sales and profits, improved cost-effectiveness, reduced selling costs, and shortened selling cycles.

Pricing Competitively

As professional mobile disc jockeys we are also emcees, interactive party hosts, event coordinators, music programmers, and audio and lighting technicians. As experienced professionals we have the ability to "read" a crowd to ensure our music and actions are in line with their desires. What's more, we can teach participation dances, and we can host contests and games. Do we warrant a premium fee? You bet we do! Just remember to always be worth your asking fee, and always give your customers even more than you have promised.

The pricing structure under which mobile DJs operate varies greatly and depends on a number of different factors. These include pricing by the competition in your market, the type of occasion you are booking, the time of year, the day or evening and location of the event, and the "extras" added to your primary rate. The average hourly rate in the industry is $100 to $300; however, a DJ's profit can sky-rocket by several hundred to several thousands of dollars per event, depending on lighting effects, dancers, party props, karaoke, and so on.

According to www.weddinggazette.com, the American national average cost of a wedding ranges from $17,000 to $31,000. Several reliable wedding publications state that the entertainment chosen for a wedding or party is responsible for 80 percent of its success. If DJs were paid commensurate with this percentage, we would average $19,200 per wedding gig! Although that is not likely to happen anytime soon, does $2,400 for a four-hour booking get your attention? That's just 10 percent of the average budget for 80 percent of its success. Although $2,400 may not hold up in your market, how about $1,200, which is currently the average, according to the American Disc Jockey Association's website information on "DJ Hiring Tips"?

Experience and reputation are two factors to consider when determining your pricing structure. A newly formed business may want to consider charging less than an established business with years of experience and a solid reputation. As your reputation and experience grow, slowly increase your rates.

If every DJ added just 10 percent to the cost of his or her existing services, the entire industry would benefit from this price increase. I am not suggesting "price fixing," but I am suggesting creating an industry standard where there is room for low-, middle-, and high-end packages. Perhaps you can bring this up at a local chapter meeting of your professional organization.

It is wise to base your prices on a three-hour minimum, with higher rates during peak periods. These include April, May, June, September, October, and December. Other peak periods include

New Year's Eve, Friday evening, and Saturday day and evening, as well.

Weddings, bar/bat mitzvahs, and some corporate events are typically booked at a higher rate than other types of affairs. It is standard to charge a higher rate than other types of bookings for functions where the entertainer has the dual responsibility of being both the DJ and the emcee.

Sometimes a good incentive for people to hire your DJ service is to offer them a discount. Build this factor into your pricing structure, and consider offering discounts to:

- Nonprofit organizations
- Customers willing to prepay for an event
- Corporate clients who book your DJ services four or more times in a calendar year (apply the discount to the last job)

You can earn additional money through playing overtime at events. On-site, about a half hour before the job is over, ask the client if he or she wants you to play overtime. The overtime rate is typically 50 percent of the hourly rate you have quoted for each thirty minutes of overtime you play.

When pricing a job, take into account not only the time you spend performing but also the time you have to spend on event planning, meetings, transportation, setup/breakdown, and administration relevant to the event. These services priced individually would be considerably more for the client than as a part of a bundled fee.

Handling Deposits and Payments

As a businessperson you must ensure that you are paid for your services. One way to do this is to always have clients sign a legally binding contract. This should be returned to you along with a deposit within ten days of the booking of your services. The deposit should range from $100 to half of the total bill.

Nearly all customers will respect your professionalism if you explain in a friendly manner that you book on a first-come, first-served basis and that no date is reserved until the deposit and contract are received. It is essential to express both over the phone and in your contract when the balance is due. Apply this policy to all clients, including friends, relatives, and co-workers. It is best to have an event paid in full by the date of the performance. Some mobiles DJs prefer to have the balance due on the date of the event prior to their start time.

In the event you receive an offer of a booking at the last minute because another company's DJ has canceled, arrange for the client to sign a contract as soon as possible and to pay by cash, money order, cashier's check, or credit card.

Most DJs require a deposit from their clients to guarantee a booking for a particular date. The amount of the deposit varies from 10 percent to 100 percent, with the average being 50 percent. There are a variety of factors to consider when establishing your event cancellation policy (See the "Entertainment Agreement" in the chapter on sample materials). As with any legal document, go over your contract with an attorney to ensure that it follows the legal requirements of the area in which your disc jockey service operates.

Organizing Clients and Gigs

There are software programs available that are designed specifically for mobile DJs to increase productivity. They feature a screen with all the fields necessary to take client information over the phone. They also include the ability to take the information from your initial data entry and print a contract. The contract form can be customized to include your terms and conditions.

Using a software program for event management reduces duplication of effort and creates a "paperless" office environment. You need only input the client's name and address once and from that information, print out an envelope, contract, cover letter, invoice, or a variety of other forms. Additional information such as payment arrangements and a client's music preferences can also be entered.

The calendar function allows you to look up a certain date to review your bookings and their status such as "pending" or "confirmed." A venue database section can contain directions to each of your regular event facilities. There is also a feature that allows you to view all pertinent client information on one screen.

Web-Based Business Tools

- DJWebmin—www.djwebmin.com
- DJ Intelligence—www.djintelligence.com
- Realtime—www.realtimedj.com

DJ Business Administration

- DJ Manager—www.djmgr.com
- DJ Calendar—www.djcalendar.com
- Infomanager—www.cwarenet.com
- DJSoft—www.djsoftinc.com

Music Database Software

- CD Trustee—www.base40.com
- Music Collector—www.musiccollectorz.com
- Music Magic—www.musicmagic4u.com
- CATraxx—www.fnprg.com/catraxx
- Music Database 2000—www.accsi.com/music/database.html
- Visitrax—www.synapsa.com

Equipment

As a professional DJ, a large part of the value you have to offer clients is based on the quality of the equipment you use. Fortunately, there are an abundance of excellent sound and lighting products and accessories available.

I am frequently being asked by my DJ peers, "Stacy, what gear do you use?" I am happy to use this forum to answer this question. Currently, I am an Artist Representative for American DJ lighting (www.americandj.com), PVDJ (www.pvdj.com) audio, and Promo Only (www.promoonly.com) CDs and DVDs.

All of these manufacturers are truly committed to the disc jockey industry. I use their gear and heartily endorse them to the reader. Here's why:

My American DJ light show brings visual excitement to the dance floor at any event. It also makes me more money as an up-sell, and through referrals. The products are reliable, affordable, and easy to use. The main effects I use are the Whirl 250, Vertigo, Quintet, and Avenger II, along with two, 4-channel lighted chasers, all mounted on an LTS-2 tripod stand.

PVDJ's DPC 1400X is a single-rack-space digital power amplifier that uses patented digital power conversion technology. The revolutionary design delivers crystal clear highs, thumping low-end response, and weighs only 15 pounds. It can operate in both stereo and bridged modes. My DJS 5 speakers are attractive, reliable, and they provide a crisp reproduction of pre-recorded music that is outstanding. The field replaceable baskets give me peace of mind while on the road. The Kosmos Audio Enhancement System gives my full-range speakers extra bass without muddying the quality of sound.

As a mobile DJ I perform at a wide array of events. To be prepared, I need the hottest new musical and visual tools. My Promo Only clean edit Mainstream and Country Radio CDs and Hot Video DVDs ensure that I don't miss a beat! The variety gives me outstanding musical programming capability so I am able to keep people dancing and having fun.

Accessories that I use and recommend include:

- SKB 106DJ Station Road Case to house my components because it is durable, weighs in at less than 13 pounds, and has rotomolded side handles. (www.skbcases.com)
- AClassActFX.com 32 inch streamer/confetti launcher. The company is the provider of the original confetti, streamer, and T-shirt launchers (cannons) and supplies. Made in the United States, the paper is FDA-approved nontoxic, flameproof, biodegradable, and bleed resistant. The company has been granted two U.S. patents for their launcher system. (www.AClassActFX.com)
- Cosco's Endura center folding table with built-in, easy-to-carry handle. It is 72 inches long × 30 inches wide × 29 inches high, weighs only 42 pounds, and holds up to 600 pounds of weight with 1-inch heavy-duty tubular steel frame

and legs. The table top is moisture proof and stain resistant. (www.coscoproducts.com)

- Camelback Display's 6-foot black table cover. It is 100% polyester, which is the best fabric for avoiding wrinkles, is machine washable and flame retardant, and is available in a variety of sizes, with or without your logo. (www.camelback-displays.com)
- Cord-Lox Coil 'n' Carry handles and Velcro fasteners. Cord-Lox makes a complete line of cable management products in a wide variety of sizes, styles, and colors for use in the professional sound and other industries. They can also fabricate any type of strap you may need. (www.cordlox.com)
- Harper Senior 2-in-1 hand truck. It has a 600-pound capacity, is made of ultra tough nylon, and features pneumatic wheels. I bought mine at Ace Hardware. (www.harpertrucks.com)
- American DJ's multipurpose gig bag. It has ample pockets for storing contracts, song lists, and loads of small items that are needed for every event. (www.americandj.com)

Here are some of the other items I use with my mobile rig:

- Furman PL-Plus Series-II Power conditioner and light module (www.furmansound.com)
- Pioneer HDJ-1000 Pro DJ headphones (www.pioneer prodj.com)
- American Audio Protek/CD300 CD roadcase (www.adjaudio.com)
- Littlite 12-inch black gooseneck lamp for mixer (www.littlite.com)
- Hosa 14-gauge, 25-foot and 50-foot black speaker cords (www.hosatech.com)
- Furman 50-foot black extension cord with three outlet female sockets (www.furmansound.com)

DJ/KJ Stan Slavik uses his graphics and marketing background for the Karaoke Connection van design.

DJ icon Bernie Howard is geared up to provide Emergency Music Support with his "Jambulance."

Marketing is not an event, but a process. It has a beginning, middle, but never an end. You improve it, perfect it, change it, even pause it. But you never stop it completely.

JAY CONRAD LEVINSON

3

Marketing: More Than Your Magnetic Personality

Advertising, public relations, media planning, pricing, sales and customer service are all parts of your comprehensive marketing effort. As you market your company and its services, keep in mind that all the facets must work together to create an effective strategy.

SCOTT FAVER

Photo courtesy Jann Gentry, Jann Gentry Photography

Owner/Entertainer: The Party Favers, California and Arizona
Member: American Disc Jockey Association, National Association of Catering Executives, Business Networking International

The Marketplace

The demand for mobile disc jockey entertainers is growing. People who would have hired a live band for previous functions are now hiring DJs for two main reasons. The first is cost. Mobile DJs are generally less expensive than bands. In addition, disc jockeys almost never take breaks, and clients get more music for their money.

The second reason is flexibility. A DJ provides all types of music. This includes Big Band, Country, Rock, Oldies, Hip-Hop, Alternative, and everything in between. Very few bands can competently cover such a wide range of music to suit every taste. Communicate to potential clients that hiring a professional mobile DJ emcee far exceeds the entertainment that a band can provide. Illustrate how by showing your company DVD.

The market for mobile DJs is varied because not all of us pursue the same market. It is important to position your company toward the type of market in which you will specialize. It is necessary to gear all your advertising and promotional literature toward this specific market or markets, depending on your specialty. In marketing, perception is everything.

A focused marketing campaign will yield far better results than a scattered one. There are several factors people consider when deciding on a DJ. They will want to know about your experience with the type of affair they are holding. In addition, they may want to discuss the range of your music and whether or not you take requests (of course you do!). They may also make inquiries about your sound system. They may ask if you offer lighting and video effects, or party-planning services.

Marketing Basics

Here are the Top 10 attributes of a highly effective marketing campaign:

1 **Profitable:** It motivates people to buy what you're offering, and makes more money than it costs.
2 **Long-lived:** It maintains its selling power and appeal over time.
3 **Unique:** It stands out from your competitors by showcasing your company's unique features and benefits.

4 **Substantive:** It places substance ahead of style, emphasizing the benefits of your services without relying on gimmickry.

5 **Client-focused:** It connects with your prospects and focuses on their needs and interests, not yours. It engages them by talking to them, instead of about you.

6 **Solution-oriented:** It clearly explains how your prospects' problems or issues can be solved, or needs and desires fulfilled, using your DJ services.

7 **Permission-based:** It involves getting people's consent to market to them, and emphasizes creating a relationship with them.

8 **Tailored:** It is specifically crafted for delivery based on the strengths of various media (Internet, *Yellow Pages*, bridal publications, etc.)

9 **Clear:** It is clear, concise, and easy to understand.

10 **Creative:** It shows distinctiveness and flare, but always with the goal of showcasing your services and getting people to book you.

To succeed and grow your disc jockey business, you must become a master marketer of mobile entertainment services. You must also view marketing as an investment, not an expense. In general, the term *marketing* covers every aspect of sales, advertising, public relations, promotion, networking, and customer service.

A large part of the success in your marketing strategy can be attributed to the consistency of your message. So repeat your message over and over again. People need to know about your company before they can contact you. Therefore it is essential to get your message out repeatedly to gain enough exposure for them to do so. Be consistent in your use of a company slogan, headline, and so on. Timing is also important. You want to reach target prospects when they are planning an event and have need of your services.

You cannot succeed in business solely on the merits of your excellent performances. Prospects who have never seen you entertain will initially judge your company based on your marketing materials. Therefore, you must create the perception of quality, reliability, professionalism, trustworthiness, and high value in the way you market your services. Only then can you attract high-caliber prospects who are willing to pay your asking price.

Effective marketing needs to be organized, dynamic, integrated, and consistent. A proper marketing strategy includes these components:

- Target market(s) identification and research
- Competition identification and research
- Budget estimates
- Marketing tools selection
- Marketing plan execution

Researching the Competition

To effectively conduct competition identification and research, you must answer the following ten questions:

- What services does a particular competitor offer?
- How many DJs work for the company?
- Where is it located, and what are the hours of operation?
- What are its sales and marketing tactics?
- What is/are its target market(s)?
- How much does it charge for different types of events?
- Are its clients satisfied?
- Do many of its gigs come from referrals?
- What percentage of its jobs are repeat customers?

Granted, the answers to some of these questions are not easily identified. "Resourceful" is the key word when it comes to conducting research on the competition in your marketplace.

Marketing's Various Facets

Here's a humorous illustration that gets right to the heart of what tools you can employ to market your skills:

- A beautiful woman at a party you're attending reads a sign that says, "Meet Tom Tonight Only! He's a great guy, and he's very rich. See bartender for more details and a free cocktail." **That's Advertising.**

- You see a beautiful woman at a party. You go up to her and say, "You are the most beautiful woman that I've ever seen, and by the way, I'm a great guy and very rich. Would you marry me?" **That's Direct Marketing.**
- You see a beautiful woman at a party. You go up to her and ask for her telephone number. The next day you call and say, "You are the most beautiful woman that I've ever seen, and by the way, I'm a great guy and very rich. Would you marry me?" **That's Telemarketing.**
- You're at a party and see a beautiful woman. You straighten your tie, buy her a cocktail, hand her a newspaper clipping about yourself, and say, "You are the most beautiful woman that I've ever seen, and by the way, I'm a great guy and very rich. Would you marry me?" **That's Public Relations.**
- You're at a party with several friends and see a beautiful woman. One of your friends goes up to her and, pointing at you, says, "He's a great guy, and he's very rich. You should marry him." **That's a Testimonial.**
- You're at a party and see a beautiful woman. She walks up to you and says, "I know who you are. You are an eligible bachelor who is a great guy and also very rich." **That's Brand Recognition.**
- You see a beautiful woman at a party. You go up to her and say, "You are the most beautiful woman that I've ever seen, and by the way, I'm a great guy and very rich. Would you marry me?" The two of you go out on a date. You make a romantic overture too quickly and she slaps you in the face. **That's Customer Feedback!**

Branding

Branding is the often subliminal process by which a business employs marketing strategies to get people to easily remember their products and services. Although logos and slogans are usually the first things that come to mind related to branding, branding is not just about them. It's the entire "feel" associated with a company, which should provide a customer with a sense of security.

Branding consistency needs to apply to all aspects of your marketing, advertising, public relations, and sales efforts, because branding establishes recall about your information.

In order to brand something, give it that special something to make it stand out from the rest. Dare to be different from your competitors in your branding by using creative graphics, logos, headlines, and slogans. My company's slogan, for example, is "Bringing the *Party* to Your Event!™"

Carving Your Niche

A common mistake among newcomers in any field of business is trying to be "all things to all people." Nothing can be less productive than this approach. Specializing and narrowing one's focus as much as possible will actually increase the likelihood of getting more business. Specialization is in itself a fundamental marketing process. It is amazingly effective in creating "top-of-mind" awareness with a specific target market.

Having a market niche enables you to differentiate your DJ service from competitors' services. Your marketing efforts should be geared toward dominating your niche market to gain the competitive advantage and bring you sustainable business profitability.

Begin the process by making a list of services, products, personnel, and capabilities that your company can offer to potential customers. The list will reveal your primary selling points. Use these features and benefits to take ownership of a market niche in a way that makes your position virtually unassailable. Specialization casts an aura of superiority and exclusivity. When people deal with a specialist, they automatically assume that this person has greater knowledge and expertise than a generalist.

Specialization is the wave of the future. The greater the competition in your region, the greater the need for more specialists to dominate segments of the market.

Permission Marketing

"Permission marketing" basically means that a prospect or client has given a business permission to make contact with specific types of offers or information. This can be done via e-mail, regular mail, or even telephone calls.

In our business, the people who usually agree to permission marketing are past clients or prospects who are interested in something you have to offer.

Telemarketing

Telemarketing—marketing your services over the telephone to pre-qualified prospects—can be an invaluable tool for initiating new business. Telemarketing is accomplished by either personally calling the prospect or hiring a service to do this for you. Clearly, it is preferable for you to make these contacts yourself, if feasible, because you are generally the best spokesperson for your business. However, if you need to contract this service out to a marketing firm, be certain you are clear on who the callers will be, the type of training they receive, the monitoring that is done to ensure quality, and the time allotted for each call. This will give you a measure of confidence that your message is being delivered in the same professional manner you would use.

The goal of your telemarketing effort is to get people to hire you, or request more information about your services. To be effective, you must begin with a targeted list of prospects and prepare a script of what you will say. Now you are ready, so pick up the phone and make those calls!

Your target prospects may include nonprofit organizations, corporate activities directors, brides-to-be (engagements are listed in the local newspapers), high school and college activity directors, newly opening retail stores (check with the local chamber of commerce), country clubs, party equipment rental stores, restaurants, night clubs, function halls, river cruise boats, hotels, and so on.

Some bridal publications offer lists of engaged couples. Make sure that any such lists are tailored to your needs before spending any of your marketing budget on them. You can find the information you need to develop these lists in the *Yellow Pages,* newspapers, and business journals.

Developing key contacts will significantly assist your telemarketing efforts. Take the time to meet with key people in person. Consider offering them a financial or other personal incentive to provide you with referrals (but be careful not to violate any of their company policies, etc., when doing this).

If you have worked out a mutually beneficial arrangement with a key prospect, do not wait for your contact to call you. The responsibility lies with you to regularly stay in touch. The ideal situation will be one in which you develop a long-term relationship. In this instance, the initial time and effort you have expended will pay off in lots of ongoing referrals.

Perhaps the most important prospect list of all is that of your past customers. Call or write to them twice yearly to inquire about their future occasions and any referrals they may have for you. You may want to offer past customers a financial incentive for helping you with referrals or for becoming a repeat client for your DJ service.

Telemarketing Scripts

An important key to a successful effort is to write a truly effective script. The script should not sound as if it is being read. Its tone should be friendly, conversational, compelling, and brief. Readers should be rehearsed enough to be able to veer from the script in response to the prospective client and then move back to the script without sounding as if they are reading. Aim the script toward the type of event the prospect may be planning. For example:

> *"Hi, Janet, this is John Peterson from Platinum DJ Service in Northampton. I understand from the newspaper that you are getting married on June 22. Is that right? Congratulations!*
>
> *The reason I'm calling is to let you know that Platinum DJ Service specializes in entertaining at wedding receptions. In fact, we were recently voted "Best Wedding DJs" by* Entertainment *magazine, and have hundreds of happy brides and grooms to thank for that.*
>
> *I am confident that we can provide you and your fiancé with all of the elements to make your dream wedding come to life. Do you have some ideas about what you'd like your reception to be like?*
>
> *Let's get together for a consultation so that you and Frank can both tell me about your visions for your special day. There's no obligation involved, and at this meeting I will present detailed information about our services, give you glowing testimonials from satisfied clients, and even show you a DVD of our wedding DJ/emcees doing events.*

Is getting together to discuss customizing your wedding reception of interest to you? Is this week or next a better time to meet? I have Tuesday evening at 6:00 P.M. or Thursday evening at 7:30 P.M. available. Which day is better for you and Frank?" And so on.

Price is often a primary concern. Even if a prospect believes you are too expensive, offer to send your brochure. You may still get the booking if, after speaking to you, the prospect calls several other DJ services and finds they cost more or they are not as friendly and appealing as you were on the telephone.

Marketing Packet

The impression you make on a potential client with your marketing packet can be the difference between getting hired or not. If you want to be viewed as a polished and experienced professional, then all your materials should convey this image of your disc jockey business. Although it may be true that "you can't judge a book by its cover," people do judge businesses by their initial presentation.

For a truly professional appearance, your marketing packet needs to be visually appealing, uniform in color and design, organized, and easy to read and understand, with *perfect* grammar and sentence structure. It should clearly state the benefits of acting immediately to contact your company about booking its DJ services.

Here is a list of materials to consider using in your marketing packet:

- Cover letter
- Brochure
- Business card
- CD-ROM or DVD
- Contract/Proposal
- Condensed music list (or website suggestions where the list can be viewed)
- Event planner
- Client references or testimonials
- Copies of a "How to Hire a DJ" article (available at numerous DJ and wedding websites)
- Quality copies of any press coverage you have received
- "Frequently asked questions" handout

- Insurance documentation
- Professional association membership certificate copies

Presentation

Two-pocket glossy folders are a great way to keep your documents organized to hand out to prospective customers. They can be purchased at any office supply store. You can personalize them by adding a label to the front, or have a printer customize them with your business logo. There are a variety of printing styles from which you can choose, including raised ink or embossing. Your best bet is to shop around for options and quotes. Many printers will give you a lowered package price if you print all your marketing materials with them at once or if you promise routine reorders over time.

Direct Mail

Your direct mail package should include a cover letter, brochure, and, in some cases, a coupon. It is important for you to know that the goal of direct mail is not to sell your service but rather to prompt a response so that you can then sell your service when you make contact with a prospect.

You can target zip (postal) codes to reach market areas that are right for your business. Some bridal publications offer direct mail lists of engaged couples. These lists are often broken down by geographic region and can be highly effective in tapping into the lucrative wedding market.

Sending two or three direct mailings per prospect will yield the greatest results because it vastly improves the odds of reaching prospects at the right time. In addition, direct mail is most effective when used as part of your overall marketing efforts.

For example, first a bride sees your ad in the wedding supplement of your local newspaper. Next, she sees it in a wedding publication. She meets you in person at a bridal show and then gets a mailing from your DJ service. By this time, she feels she has known about you for a long time and is sure you are highly regarded, because you are featured in so many quality media that she is relying on to plan her event. Cross-media advertising combined with direct-mail marketing is a truly powerful combination of marketing techniques!

In your efforts to reach the right target market, you may want to obtain listings such as getting the names of area high schools and colleges, to market to their student publications and organizations. You can find lists of people getting married, and organizations and businesses that regularly put on social events. All of these are a good way to start the mailing and calling process.

In many cases you will only be able to get a general list so you'll need to do additional research before sending out your mailing. For example, you may have the name and address of a school or business, but no contact name. Always send your mailing to a specific person. This will greatly increase the chances of your information being read and retained by the person who is making the decision. If possible, always follow up on a mailing with a phone call within the week. You may want to say that you are calling to make sure the person got the information you sent and to ask if there are any questions. Inquire about upcoming events. This person is your key contact, so "schmooze" a little.

With schools, send out separate mailings to the class presidents of each grade level as well as to the class advisers, student council presidents, student activities directors, and prom committee chairs.

As for wedding business development, you may want to hold a contest at a bridal fair. Once you have a list of brides-to-be, send mailings to them. Also send mailings to engaged couples whose names are announced in the newspapers. Other important direct mail lists will include the hotels, banquet facilities, and restaurants that host parties that are located within the geographic area you serve. Folks seeking a mobile DJ entertainer will often call these places and ask for a referral. You want your information to go to the person who is in charge of bookings. Make sure your brochures and business cards are in the hands of every wedding-planning service and event-planning company in the region, too. Here are some other strategies:

- Ask the people and vendors with whom you spend money to recommend your services. Offer incentives.
- To become a one-stop entertainment shop, add more "profit centers" to your business. (e.g., offer party props, lighting, photography, video, and so on)
- To reduce overhead, subrent space in your existing location to another professional or seek out a successful noncompeting service from which you can sublet.

Letterhead and Envelopes

Your DJ company will need high-quality bond letterhead and envelopes that coordinate with your business cards. It is okay to be a little artsy or unusual in the design. Just try to stay within accepted professional limits. Remember, the point is to please and attract clients, not dismay or puzzle them. It is important to maintain consistency of image with all of your company's business forms, communications, and marketing materials.

Business Cards

Think of your business card as a miniature billboard. Business cards are your single most important marketing tool and create a lasting impression. Therefore, they should be professionally designed and visually attractive. You will know you have accomplished this goal if/when you hand your business card to someone who says, "Wow, great card!"

The business cards for your DJ service should include a slogan. "You have the dream. We make it happen!" is one such example, used by The Party Favers. It can also include a positioning statement such as "Disc Jockey Entertainment for All Occasions. Bar/Bat Mitzvahs a Specialty;" Your name and title; a professionally designed logo for your business; telephone, cell phone, and fax numbers; e-mail and website addresses are all necessary to ensure that customers can reach you easily. For security purposes, do not include a home address; use a post office box or give no address at all. It is also preferable to use a business telephone line, rather than a home telephone, if at all possible. This way, as your business grows, you will be able to establish a definite division between home and work. If you operate out of a commercial location, then do include that address and phone number.

Spending a little extra to have all of your business materials look highly professional is well worth the expense. Take your ideas to a designer to have materials designed to go together. Have good-quality printing done, and use good paper. Splurge a little on color if you can. (See Sample Materials for business card samples.)

To include ample information, consider using both the front and back of your business card. To make them stand out, have the cards printed with colored ink or on colored paper. Gold or silver foil stamping on shiny black stock is a nice alternative to the standard white stock. You may want your business cards printed on extra thick stock so they are weightier than most. UV-coated cards are fast becoming the business card of preference in our industry.

Another important expense to consider is investing in a professionally designed logo to create your mark of distinction. Logos from graphic designers cost between $200 and $2,000 and up. Although this can certainly be a stretch for a new business, it is worth either having it done or waiting to produce your final materials until you can afford to do this.

Distribution

At gigs, place a small stack of business cards in a holder on top of the table where your sound system is set up. If you are using subwoofers, put some on top of them, as well. Also place some cards near the tip jar at the bar. Of course, always carry some in your jacket pocket. When someone compliments you on a job well done, be proactive. Respond by saying, "Thank you, here's my business card."

Be sure to swap cards with vendors at events. Give a few of yours to the banquet manager, caterer, photographer, videographer, and bartenders. Ask these folks for their cards, as well. With luck, they will be impressed with your talent and pass the cards on to potential clients.

Here are some other places you can leave business cards to "drum up" prospects:

- Bridal shops
- Community and school bulletin boards
- Party supply stores
- Music stores
- Organizations to which you belong
- Retail businesses

At any retail business where you are a customer, ask if you can leave several cards on the counter. Be sure to hand out your business cards to neighbors and co-workers if you work outside the DJ industry.

If it fits into your marketing budget, you might also consider producing a digital CD-ROM business card. This combines a conventional card with the data storage of a CD-ROM, and can be played in any standard CD-ROM drive or CD player that handles the 3-inch size.

Your DJ company's digital business card should have all pertinent contact information on the top, printed in color. This medium presents a high-tech format to distribute brochures, photographs, audio, and video clips. It can even be used to promote your website with links to its online content. Of all the forms of business presentation you have available, this is one of the most effective in making a lasting impression and getting across a complete message.

Brochures

Having a brochure to offer prospective clients is helpful to your marketing and sales efforts. Desktop publishing has provided a breakthrough service for small business owners: There are now preprinted brochure templates that can be purchased and used with a publishing software program to create professional-looking promotional literature. You can also have your brochures professionally designed and printed on slick paper with multiple colored inks. However, be advised that there is a large cost difference between these two methods.

If copy writing and design isn't in the mix of your talents, then absolutely hire a professional to design your brochure. It is not advisable or effective to invest good money in promotional literature that looks amateurish in content and design.

Consider creating three-fold, 8½- × 11-inch brochures, or two-fold, 16- × 11-inch ones. Effective brochures have the following qualities:

- They use two to four colors.
- Pictures and/or graphics are included.
- There is a lot of "white space" around the copy.
- They sell the benefits of your services, as well as allay the readers' fears and appeal to their emotions.

The image presented to potential customers through your promotional literature needs to be top-notch, because *perception is reality.* The more professional you appear to be, the quicker you will move to the forefront of the professional DJs in the region. Therefore, having a polished and professional brochure is worth every penny of the expense.

Cover Letters

Statistics show that the most effective cover letters used for marketing contain the following qualities:

- They are personalized—addressed to a specific person and signed by the business owner or a high-ranking member of your company.
- They are two-sided and are printed in either Times Roman or Arial typeface.
- They contain headings and subheadings printed in blue.
- They are written in conversational language.
- The first side ends in the middle of a sentence (prompting the reader to turn the page)
- The margins are not justified (lines are uneven length).
- Key points are underlined.
- They offer a risk-free proposition.
- There is a "magnetic" P.S., or add-on, at the bottom of the letter that emphasizes your offer.

You can choose to vary from this formula somewhat, but do not go too far from it, or you will risk spending a lot of time and effort on a mailing that may not meet your expectations.

Newsletters and E-zines

Newsletters and e-zines can be used as an excellent marketing and public relations vehicle to increase business. The key to success is consistency. They can be sent to both current and potential clients for holidays and/or birthdays, or on a quarterly basis.

If wedding receptions are your DJ service's specialty, consider producing a targeted newsletter or e-zine for engaged couples. This can be used to provide helpful suggestions that span across all planning phases of a wedding—from choosing vendors to suggested seating arrangements. E-zines and newsletters can be e-mailed to clients and prospects. Hard copies can be handed out at bridal fairs.

Your mailing list can include couples who have had engagement announcements in your local newspaper, as well as bridal consultants, formal wear shops, catering and banquet facilities, and restaurants in your area. Be sure to include testimonials from past wedding clients in the mailing, as they likely have friends who can benefit from your entertainment services.

Here are some ideas for information you can include in your newsletter or e-zine:

- Reasons why the DJ should be booked immediately after the venue is secured
- Popular specialty dance songs
- A listing of reception format options
- How to deal with wedding vendors
- References from past clients
- Ethnic traditions
- Wedding games
- Budgeting your reception
- Wedding websites

The list can grow and grow as you think of areas in your business you would like to cross-promote while filling the pages. Whatever the content you choose, make your publication stand out. Use easy-to-read fonts and fun illustrations that add *pizzazz* and flair to your presentation. Do not overcrowd words on a page, and keep enough white space for easy reading. Most of all, grab readers' attention with powerful and intriguing headlines that make them want to keep reading.

Free newsletter templates are available at www.microsoft.com or in many desktop publishing packages. Many others can be located for free by browsing around the Internet. Depending on your technical abilities, M.S. Publisher, Quark, and InDesign are popular software programs in which you can design your own hard copy newsletter.

Mailing the newsletter is the cost of a stamp. If you use an $8^{1}/_{2}$- × 11-inch or 11- × 17-inch format, you can eliminate the need for an envelope by folding the publication in half. Just leave the bottom half blank for the addressees and your return address.

Instead of stapling the two halves together, use adhesive dots, which can be purchased at any office supply store. If the quantity of newsletters that you plan to mail justifies the cost, consider applying for a bulk mail permit.

Websites

Launching and maintaining a website for your business is easy. There are low-cost services available to help you turn out a site that will allow people to learn about your products, book your services, and pay for them all at the same time. A business without a website is at a distinct disadvantage over one with a site. When designing and planning your site, keep its visual image consistent with your other marketing materials.

Production

The most important aspect of producing a website is the planning. The purpose of your site should be for the visitor to want to hire you or request a phone call or appointment. Make it easy for them!

When producing a website, the primary areas to consider are navigation (how the client will get from one page to the next); content (the information in the site itself); maintenance (how you will keep the information updated), mechanics (where your site will appear on the Internet and how much it will cost), and marketing (how people will find out about your site). Continuity is important, so if you have developed a certain look or style for your DJ service, develop a website that is compatible and reflects that image.

Your home page should include the headings on your site, your logo, a few photos, and links to your other pages. Include a separate page for each of your company's divisions (such as weddings, schools, mitzvahs, and so on) with action photos. Your website should contain the names and pictures of your staff DJs with links to their bios. Your music library database is another feature you can add. Be sure your company profile is in the mix.

Your site needs to be registered with the larger search engines, which will allow people to find you. There are both general-purpose search engines and specialty search engines (such as for disc jockeys, brides, etc.).

Whether you register your own address and host it with an Internet service provider (www.mydjservice.com) or sublease website space (prodj.com/mydjservice), you have made an investment in your Internet presence. It is like graduating from a free listing in the Yellow Pages to a display advertisement.

Major features on "produced" sites include online music databases that are searchable by both title and artist and forms that let clients complete a Wedding Reception Planner online. DJ Intelligence (www.djintelligence.com) offers online interactive tools, ready-made for your disc jockey website.

Features

A lot of technically sophisticated features can be incorporated into a website. These include the options previously noted as well as being able to check your availability, online payments, and form processing.

A links page with local vendors is a great way to network. Some high-tech options include a price quote generator, animated graphics, buttons and menus, banners, bullets, backgrounds, animations, tables, frames, sound, and streaming video of your DJs in action at different types of events. Using all of the latest multimedia features is like printing a 3D, interactive brochure. Easy navigation around your site is the key to giving clients the information they want fast.

Content

A good starting point for writing website copy is your existing company brochure. It is a relatively easy task to put brochure information and photos on your website. This will help your website developer to understand your business and its direction.

The visual look, content, and graphics in your DJ service's website should be based on your target market(s) and goals. Here are some examples of goals:

- Inform prospective customers about your company's services.
- Provide specific product details and specifications.
- Generate and close sales.
- Provide a forum for communication with customers or prospects.

- Enhance your reputation through the use of this new technology.
- Provide an easy way to download music lists.

Unlike a brochure, which limits the amount of copy you can fit, your website can hold all the copy you can generate. Here are some things you can include in your website:

- **Press Releases:** Add your company press releases to a page, so prospects can read about your latest developments. This also legitimizes your business and gives a prospect a sense of confidence that you are a responsible business owner.
- **What's New?** A "What's New?" section is a page that has links to the latest products or services you have. Research shows that What's New? pages are heavily viewed.
- **Links to Other Websites:** Actually, you might not want to provide links to other websites, although many companies do. But consider this. Why would you want to make it very easy for someone to leave your site? What if the site that person links to has information on your competitors? This one is a judgment call, so consider it carefully. In some cases, you may want to provide links to industry groups such as the American Disc Jockey Association (www.adja.com). Whatever choice you make, try to make it hard to navigate to the links page until most of your site has been viewed. For instance, the links button can be placed only at the end of the "Contact Us" page.
- **Employment Opportunities:** Offering employment information can be an inexpensive way of finding good personnel. But be sure to keep the employment information well separated from customer information.

Cost

There are several costs associated with running a website. They include site development, hosting, and maintenance. The cost of producing a website ranges from a few hundred to several thousand dollars depending on the complexity of the site and the features it has available.

To cut down on costs, consider hiring a company that will bundle all the cost of design and maintenance into a consistent

monthly fee. A good company will design the site so that maintenance costs are minimal. When planning your marketing budget,
be sure to include the cost of registering your site with the various
search engines that will bring it up when keywords are entered in
the search box.

Maintenance

Keep your site current by revising existing web documents or creating new ones. You may do this yourself, or revising could be
included in your hosting package. Another idea is to outsource revi-
sion work. Your choice is best based on your budget, time, and skills.

Customer Service

Check your e-mail at least once daily, and respond promptly. People
expect e-mails to be answered within hours; a day at the most.
E-mail should be treated much more like a phone call than like
regular mail. If you travel, you need to arrange remote access or
have an associate check for you. Create an automatic response to
answer every e-mail with a message that says the client is important, the client's message is received, and you will respond shortly.
If you fail to respond quickly, other DJs will, and you will be off the
short list of service providers.

Marketing

To ensure the greatest success of your site, you will need to actively
market it. Be sure to put your website address on your business
cards, letterhead, brochures, company vehicles, *Yellow Pages*, and
other advertisements, as well as your business answering machine
and fax and your company's promotional items.

Inform prospects of your website address on the telephone or
during in-person consultations. Place it on web directories and
search engines. Have your home page linked to other complementary pages on the web such as with photography and video services
with whom you may be affiliated. Research other DJ's well-
produced websites that quickly show up in the search engine
searches (e.g., Philadelphia Wedding DJ). Find out what they are
doing, and follow their lead. Pick out sites you most admire, and
send links to your designer while you plan your site.

Some disc jockey associations have websites with directory listings of their members. Many will provide you with a free link as a benefit of membership. Another option is to pay for link sites that will market your company to prospects for a fee. Other websites can also be found that offer free listings that include a link. One such site is www.partypros.com.

The Web is in a constant state of change and improvement. Almost daily there are new technologies going online that have the potential to improve your site. With a good website that is properly marketed, you should expect to get qualified sales leads that lead to bookings and revenue for your business.

*When dealing with people,
remember you are not
dealing with creatures of
logic, but creatures of
emotion.*

DALE CARNEGIE

Sales: Dollar$ from Sense

4

Learning to be an effective salesperson is the hardest and most important aspect of our business. Would you rather mix at home for your own enjoyment, or entertain the multitudes and rake in the cash? If your answer is the latter, then make a commitment to continually polish your sales skills and watch the events start adding up!

CHUCK "THE DJ" LEHNHARD

Owner: Maui Mobile Music, Maui, Hawaii
Recipient: "Mobile DJ of the Month" honor by *DJ Times* magazine

Converting Prospects into Clients

If you want to earn top dollar as an entertainer, then learning to do sales successfully is imperative. There are many resources that can help you become a better salesperson. Finding and following them are the keys to improvement.

People rarely buy products or services for logical reasons. They buy for the emotional reward it gives them, and look for logical reasons to justify their purchase later. People also like to do business with individuals they like and trust, because it provides them with peace of mind.

Several forces affect buying decisions. Primary among them are services and products that:

- Provide solutions to problems that people have or fear happening
- Represent the means of attaining their desires
- Offer confidence in the quality of the solution
- Prospects perceive as offering a value greater than the cost

In the sales process, potential customers are called "prospects." Only after you are hired do they become "clients." Your goal in advertising, marketing, and public relations is to convert prospects into clients.

The best types of prospects are people who have already seen you perform and those that come from referrals. It's much easier to close a sale when someone is already predisposed to hire you.

If your website has a price quote generator feature and the prospect contacts you to book your services, you will likely not have to haggle over prices. This type of prospect is also among the most favorable.

People invariably want to know your price before they understand the benefits of your services; however, quoting prices is meaningless until potential clients can put the cost into the context of the benefits they will receive. Allow prospects to first define their expectations, and then they'll be far more likely to sell themselves when you finally explain your pricing toward the end of a conversation.

Before spending a great deal of time talking with a prospect about your DJ services, qualify them by asking a few key questions (see the section titled "Telephone Sales" for more detail). There is no need to sell your services unless you have the date available, expertise in the type of event requested, and the person you are speaking with has the authority to hire you (and can afford your services).

If the date is not available, express regret and recommend a "friendly competitor." If there is another decision-maker involved, try to set up a conversation with all appropriate parties. If the caller does meet the prequalifications, then be ready to make an effective telephone presentation that will lead to further negotiations.

If you have correctly priced your DJ services and are selling effectively, you should convert 60% to 75% of prospect inquiries into bookings. If your closing percentage is higher than this amount, your prices are too low. If you are closing less than this percentage, either your prices are too high or your sales skills need improvement.

The Art of Upselling

There are a number of profitable products and services you can upsell from your DJ service. These options enhance your ability to serve your customers, as well as allowing you to gain greater profits with your bookings.

Upselling is a proven technique you can use to get more money at the point of sale. Offer clients who are booking with you the opportunity to upgrade to a better "package." The beauty of this is that it can increase your average sale without creating any additional advertising expense.

One way to make your service very attractive to a prospective client is to offer a complete party-planning service. You can be the client's "one-stop party shop." To do this effectively, you will need to establish business relationships with printers, limo services, caterers, photographers, florists, party goods suppliers, and other services required for typical events and bar/bat mitzvahs.

Here are some profitable additions to consider:

- Brush T-shirt painters
- Big-screen video
- Candy makers
- Caricature artists
- Carnival games

- Casino equipment
- Celebrity impersonators
- Clowns
- Comedians
- Fortune tellers
- Games
- Game shows
- High-energy professional dancers
- Hypnotists
- Inflatables
- Invitations
- Karaoke
- Light shows
- Magicians
- Props, prizes, favors, and giveaways
- Sound reinforcement
- Streamer launchers

Initially, you may want to subcontract services in which you do not have expertise, or to lease products that you do not own. Eventually you can bring these services in-house for even greater profits.

An attorney can advise you on the legal details of these affiliations; generally, you should receive a substantial discount from a proprietor for booking its product or service through your company. Try to arrange a deal in which the client pays you directly and you pay the proprietor within 30 to 60 days after receiving the proprietor's bill.

Telephone Sales

Whenever feasible, attempt to meet with prospective clients in person. But when contracting over the telephone, it is helpful to create a script to follow so that there is a logical flow to the conversation and no pertinent information is missed. Here is an example of a typical booking inquiry over the phone:

Ring-Ring.
DJ: Good afternoon, Bob Sapphire Productions.

PROSPECT: Hi. I'm planning a wedding, and I'm calling around for prices of DJs. How much do you charge?

DJ: Thanks for calling. I have party packages that range from $795 to $1,250 depending on a variety of factors. My name is Bob. May I ask your first name?

PROSPECT: Anita.

DJ: Thanks, Anita. If you'll tell me the date of your event, I'll check on our availability.

PROSPECT: Saturday, March 18th of this year.

DJ: *(After checking).* I see that I do have that date available. Anita, I'm curious, how did you hear about me?

PROSPECT: I saw your listing on the www.abcdj website. *(Allows you to track your advertising and marketing efforts)*

DJ: I see. Have you spoken with other DJs?
(The answer will give you an indication about whether or not she is a price shopper.)

PROSPECT: Only one other, but I plan to call a few more after you.

DJ: Okay, Anita. I appreciate your letting me know that. Are you the sole decision maker for the event, or is your fiancé also involved?
(If the prospect is the sole decision maker, then you are selling to the right person. If he or she is not, find out who is and inquire if there could be a time when you can all get together by phone.)

PROSPECT: My fiancé, Jim, said he'd go along with what I decide.

DJ: Okay. Do you have a budget set for entertainment?

PROSPECT: No, not really. The first DJ I called said he'd do my party for $495.

DJ: Fair enough. Let me tell you briefly about my experience. I am a professional DJ entertainer who has performed at hundreds of functions over the past ten years, including many wedding receptions. As your DJ/MC, I provide nonstop music entertainment for you and your guests. I come dressed in appropriate formal or semiformal clothes and can set up an hour before your reception starts. I also coordinate all aspects of the function with the banquet manager to make sure everyone is on the same page and your reception runs flawlessly.

PROSPECT: Oh, that's good, because I want to make sure everything goes perfectly.

DJ: I totally understand, Anita. Where will the party be held?

(The answer will reveal the distance you have to travel to get to the event.)

PROSPECT: At La Renaissance in Maynard.

DJ: Good choice. I've performed there many times, and the staff is top notch. What room will the reception be held in?

PROSPECT: The Acorn Room.

DJ: Nice. How many guests do you expect to attend?

(The answer will reveal the size of the room and the equipment you need to bring, as well as giving you a general idea of the prospect's budget.)

PROSPECT: About 150.

DJ: Anita, when does the event start, and how many hours of music will you need?

(The answer will reveal how early you need to set up your gear if the event starts before the DJ does.)

PROSPECT: The cocktail hour starts at 6 P.M. and that's when I'd like the DJ to start. We have the room until 11 P.M.

DJ: I'm happy to play background music for the cocktail hour and during dinner. The music I'd play during these times is kept at low volume, which is conducive to conversation. What type of music do you and your guests like to dance to?

PROSPECT: There will be a range of ages attending, so I'd like to have slow songs, Disco, Big Band, Motown, with a little bit of Hip-Hop later in the reception.

DJ: That sounds like a good mix to me. I have thousands of songs on CD that include music of various genres from the 1940s through the present. I gladly take on-the-spot requests from event guests, and my experience has taught me how to program music that will get people dancing. Thank you for answering my questions, Anita. Based on what you have told me, the fee for my emcee and DJ entertainer services is $895. This includes musical entertainment from 6 P.M. to 11 P.M., a professional-quality sound system with CD music library. And there are no additional charges for travel or setup. It also includes free event-planning services and the peace of mind that comes with knowing you've hired an experienced entertainer who offers a 100% guaranteed good time or your money back. Does what I've described fit in with your idea of the perfect wedding reception?

PROSPECT: Yes!

DJ: Wonderful! On a scale of 1 to 10, where 1 is "I want to get off of the phone right now" and 10 is "where do I sign?" what number are you at right now?

PROSPECT: About an 8. *(If 7 or higher, proceed forward.)*

DJ: Great, Anita! I'm really looking forward to helping you plan your wedding reception. We just need to take care of some simple paperwork to get the ball rolling. I'll be sending you an Entertainment Agreement that includes everything in writing that I've told you on the phone. I offer two payment options. The first is a 5% discount off the total fee for paying in full up front. The second option is a 50% deposit to reserve your date and the balance is not due until your event. Which option do you prefer, Anita?

PROSPECT: Probably the second, but how can I be sure of your quality and reliability?

DJ: Along with the Entertainment Agreement I'm sending you, I'll include a Wedding Reception Planner and a list of references from satisfied customers.

PROSPECT: That sounds good. Would it be possible to see you performing elsewhere before I sign?

DJ: I'm sorry, but my bookings are private affair. I'm happy to send you a Wedding DVD that includes all of the specialty events from weddings that I've done in the past year or so?

PROSPECT: Oh, that would do just fine.

DJ: Great. Then I'll get the materials out in the mail to you today. Please be sure to send the Entertainment Agreement back to me along with your deposit within ten days because I am holding a tentative reservation for you until that time. Okay?

PROSPECT: Okay.

DJ: Anita, before I get your mailing and contact information, I'd like to make a suggestion that will seriously up the "fun factor" at your reception party.

PROSPECT: What's that?

DJ: Well, over the years I've found that inside every adult is a little kid looking for a socially acceptable reason to come on out and play. Do you know what I mean?

PROSPECT: Yeah, I sure do.

DJ: Well, a great way to do this is to add a colorful light show that will turn the dance floor at The Acorn Room into a real party atmosphere. I have a streamer launcher that's perfect for the moment when you and Jim are introduced during a high-energy song. What do you think about these entertainment extras?

PROSPECT: They do sound like fun. How much extra are they?

DJ: The deluxe light show is only an additional $195. It includes vibrantly colored high-tech effects that move to the beat of the music. The stream launcher is $50 for a single blast, $75 for two blasts, or $99 for three blasts.

PROSPECT: Okay, yeah. That sounds great. Go ahead and add the light show and two blasts from the streamer launcher.

DJ: I'm sure you, Jim, and your guests are going to love it. Now I just need some information from you. Anita, what is your last name and your street address?

PROSPECT: Anita Goodtime. 123 I Love DJs Lane, Maynard, MA 01068

DJ: And your phone numbers and e-mail address?

PROSPECT: My work phone number is (413) 000-0000. My home phone number is (413) 000-0000, and my cell phone number is (413) 000-0000. My e-mail address is anitawantohireyou@internet.com.

DJ: Anita, thanks for calling. I'm confident you'll be very pleased with choosing me as your DJ entertainer for your wedding reception and you can rest assured that I'll do everything possible to make it a wonderful event to remember always.

PROSPECT: Thanks, Bob, for your time and all of the information.

Stacy Zemon practicing her telephone sales skills at a very early age.

DJ: You're welcome. Have a great day and be looking for my information in the mail in a just a few days. Bye-bye.

PROSPECT: Bye.

Handling Objections

Even spending time on the phone with you must return something of value to the customer. You must initially and continually earn the right to have the customer invest time and money with you.

If your prospective client needs to "think" about engaging your services, then there is a strong possibility that you have not adequately addressed his or her objections. An essential part of your job as a salesperson is to scope out what those objections are and address them to the prospect's satisfaction to obtain a booking.

Here are some quick answers to common objections:

PROSPECT: "I want to check out other DJ companies before I make a decision."

DJ: "I'll give you a 30-day money-back guarantee in writing that states if you are able to find another DJ service equal or better to ours, we'll refund your money."

PROSPECT: "I don't want information. I only want to know your price."

DJ: "We have four packages: $500, $650, $800, and $1200. Which would you like to hear about?"

PROSPECT: "How can I be sure of your quality and reliability?"

DJ: "Here are three references of recent satisfied customers.

PROSPECT: "I really don't want to answer any more questions until I know your price."

DJ: My rate for the type of event and specifications you described is $1,200. This includes up to five hours of nonstop DJ/MC entertainment, a professional-quality sound system, a CD music library of various genres containing thousands of songs from the 1940s to the present. Your event also includes free event-planning services and the peace of mind that comes with knowing you've hired an experienced entertainer who will help make your affair an event to remember. I offer a 100 percent guaranteed good time or your money back. We offer our customers easy payment through Visa, MasterCard, or Discover credit cards, or by check or cash. How does all this sound to you?"

Telephone Etiquette

How do you sound when you answer a telephone? Cheerful? Enthusiastic? Every call from a prospect can be a call from a future client. How potential clients perceive your attitude and professionalism in the first minute of the conversation usually determines whether or not they will book your services.

One of the disadvantages of the telephone is that you can't see people's facial expressions or body language. Voice inflections and tonality can, however, tell you something about their emotions. Your words, voice inflection, and tonality should come across to the prospect in a manner that he or she experiences you as an enthusiastic professional who is sincerely interested in assisting them. Your voice on the phone should also give them some of the flavor of how you will sound when performing at the party (though, of course, not at party volume).

You may find that you speak on the phone better while standing or walking. Getting up can help make you sound more animated. This comes across as enthusiasm. Deliver your pitch naturally, like you are telling a great story. Keep it fresh and not "canned" sounding.

One way to keep someone talking is to ask open-ended questions. Remember, the telephone is not a microphone; it has a hearing piece. Remember to use this end of the phone the most.

DJs with daytime jobs are forced to rely on answering machines or services. A phone-answering service is preferable, because when a live person answers the phone the rate of hang-ups is significantly reduced. When you speak with a prospect and convert her or him into a client, then you have just justified the $50 to $100 per month cost of the answering service.

If you use your home phone as your business phone as well, the outgoing message should make that fact invisible to a prospect. Here's an example of a professional message:

> Hi, you've reached Jan Sapphire, DJ entertainer. I'm not available to answer your call right now but it is very important to me. So, at the sound of the tone, please leave your name, telephone number, a message, and the date of your event, and I'll return your call very shortly. If this is an entertainment emergency, you can try my cell phone at area code (000) 000-0000. Thanks again for calling and have a great day!

If you do not use an answering service, make sure your telephone is answered only by you or members of your household who can properly take a message or are informed enough to answer some of the client's questions. Letting a child or someone who has no knowledge of telephone etiquette answer your calls is as bad as getting no call at all. By now, you've invested quite a few dollars to get the phone to ring, so make sure it is answered properly.

In-Person Appointments

In-person meetings are often required to sell wedding, bar and bat mitzvah, and corporate prospects. In this case, gear your telephone sales efforts toward getting the potential client to meet with you, not closing the sale over the phone. Once you meet with the interested party, you can close the sale in the same manner you usually close over the phone, with the added assistance of your company video and marketing materials.

With wedding prospects, make an appointment in which both the bride and groom can attend. When making an appointment with bar or bat mitzvah prospects, both parents and the celebrant will need to attend. With a corporate prospect, whenever possible make sure you meet with all the decision makers at an appointment.

Here are some tips to maximize the effectiveness of in-person appointments:

1. **Sell on value, not price:** A prospect is not paying $800 for four hours of DJ services at a party. He or she is investing this amount in exceptional entertainment and quality assurance that comes with booking your services. The $800 includes the hours you spend in preplanning consultations, travel, equipment setup, and entertaining the prospect's guests.

2. **Appeal to a prospect's emotions (both positive and negative):** Prospects fear having an event ruined by an inexperienced DJ who arrives late, in improper attire, has lousy equipment, and a small music selection. Prospects seek the emotional rewards of fun, joy, and happiness. They want to know they made the right decision and that others will validate their choice of entertainment.

3. **Use an assumptive closing technique:** All of your words and actions should convey a sense of confidence that you are the right choice of entertainers and should assume that you will get the booking.

Giving prospective customers a sense of security is a large factor in the success of your sales efforts. Provide compelling reasons to hire your service, by offering backup equipment, liability and property insurance, membership in a professional DJ organization, client testimonials, free party-planning services, and a moneyback guarantee.

People tend to procrastinate after they decide to buy something. As time passes, other things distract them, and they can forget about you. You can avoid losing many of these sales by rewarding customers for taking immediate action and by penalizing them if they do not. Give them a compelling reason to accept your offer within a short time—or forfeit the benefit of it. For example, offer a special discount price or a special bonus for booking you before a specific deadline.

When making a purchasing decision, people want to be sure they're going to get the absolute best return on their investment. Anyone can claim something generates great results. And they often do make such claims. Providing authentic testimonials and a moneyback guarantee are great ways to dissolve skepticism in a prospect.

The greatest source of continued revenue and the fastest way to increase profits is by retaining repeat customers. All promotions should be aimed, right from the beginning, at not only attracting new clients, but at keeping them around. Obtaining new clients is the most expensive aspect of business. Continued selling to existing ones is the most profitable.

Existing customers are the most important sources of new bookings through the business they generate by word of mouth. So repeat clients increase the value of your business in current bookings and in future ones as well. With a strong roster of existing clients, you can then afford to put more energy and assets into attracting new customers.

Sales DVD or CD-ROM

A polished DVD or CD-ROM of your company's disc jockeys in action can greatly enhance your ability to make a sale to potential clients who view it. The video needs to include all pertinent information about your service, the types of events you do, and any special features or packages you offer. Wedding, school, and mitzvah DVDs or CD-ROMs are essential tools for your business and are well worth the investment in having them produced.

You may consider having a different DVD or CD-ROM for each type of function you perform. A senior class adviser will not necessarily be impressed by a wedding video, nor would a bride or groom be "wowed" by watching an outrageous prom video with lots of Hip-Hop and alternative music. Match the appropriate video to the client and the occasion.

An enterprising way to get a wedding or mitzvah video made is to barter with a videographer where you are booked. Ask him or her to videotape the important parts of your performance. This person is already being paid by the client, so the charge to you should be minimal. Of course, you can offer a trade deal in which you recommend the videographer to your clients. Make sure the quality she or he turns out is to your standards.

Strong consideration should be given to the creation of this type of company audiovisual tool. Be sure it is professionally produced and edited, and no longer than 15 minutes. Providing prospects with some form of audio-visual presentation nearly eliminates the chance that they will want to come see you perform at an event.

Compensation

Fair compensation for DJs ranges from $15 to $300 an hour or more, and is usually based on the following factors:

For Solo Operators

- The price the marketplace will bear
- The costs involved with doing an event
- Experience and ability
- Event type
- How much you value your skills, talent, and time.

For Multisystem Operators Who Are Paying Their DJs:

- The price the marketplace will bear
- The costs involved with doing an event
- Event type
- The DJ's experience and ability
- Whether or not the DJ is using his or her own equipment
- Length of employment
- Specific request from a client

- DJ employee or subcontractor booked job for company
- X number of consecutive "excellent" ratings on client evaluations

A competent and likable DJ will attract far more business than a highly skilled but socially deficient one. Why? Because to a great extent, selling ourselves is a popularity contest. So, prospects are more likely to choose you as their DJ if they like—and trust—you.

Tips

Many clients do not know the protocol for tipping their DJ. Therefore it is recommended that you add the following clause to your contract: "Gratuities given to your DJ Entertainer are made at the Client's sole discretion. 10 percent to 20 percent of the total fee is customary for an excellent performance." This will also motivate your DJs to be their best because they understand that their tip is riding on their performance at an event.

The aim of marketing is to know and understand the customer so well the product or service fits him and sells itself.

PETER F. DRUCKER

5

Specialty Markets: Bull's-Eye, You're Right on Target

Carve out a niche to become the most sought-after wedding, mitzvah, or school DJ company in your market. Make a conscious decision to provide the best quality service. Use and track client evaluations for improvement. Make sure your customers understand the value of what you offer—and that it doesn't come cheap.

RANDY BARTLETT

Photo courtesy
Randy Santee,
Santees Designer
Images

Owner: Premier Entertainment, Sacramento, CA.
Producer: The 1% Solution DVD Series and Newsletter
Keynote Speaker: Association for Wedding Professionals International
Speaker: Mobile Beat DJ Show & Conference, DJ Times International DJ Expo, MAPDJ
Member: American Disc Jockey Association, Association for Wedding Professionals International

Events Galore

There are an abundance of private, corporate, and public events that are ideal bookings for a DJ. Depending on your professional skills and the skills of those who work for you, some or all of these will be right for your company.

American mobile DJs are exempt from paying licensing fees, provided that they play only for private parties or at facilities that are licensed by organizations such as ASCAP and BMI. These organizations protect the rights of writers, performers, and publishers of recorded music. They ensure that their members receive proper compensation for their work. Radio and television stations, and public entities that provide musical entertainment, must pay licensing fees. It may be wise to include a clause in your client contract that legally covers you regarding this issue. Most countries have their own laws about this. Be certain to find out and follow the laws that apply in your area.

Other than running your own dance and selling tickets, here are some of the booking possibilities that exist in the marketplace:

- Anniversary parties
- Banquets
- Bar and bat mitzvahs
- Birthday parties
- Block parties
- Bowling alleys
- Carnivals
- Clubs and organizations
- Cocktail parties
- College dances
- Conventions
- Corporate events
- Cruises
- Fairs
- Fashion shows
- Fundraisers
- Graduations
- Grand openings
- Holiday parties

- House parties
- Junior and senior high school dances
- Karaoke
- Member parties
- Nightclubs and bars
- Picnics
- Pool parties
- Proms
- Rallies
- Restaurants
- Retirement parties
- Reunions
- Singles dances
- Skating rinks
- Socials
- Teen dances
- Theme nights
- Weddings

Carving Your Niche

Niche marketing for DJs is the opposite of general marketing. It targets specific types of events that require specialized knowledge and skills.

Not all DJs have the ability to perform equally well at different types of occasions. A high school prom calls for an entirely different kind of music and entertainment style than does a dance for 35- to 60-year-olds at a country club.

Ask yourself the following questions: Can I relate equally well to both audiences? Do I enjoy Hip-Hop and alternative music as much as Motown and Standards?

Your answers to these kinds of questions will determine the types of occasions that you may choose as your specialties. It will also help you determine what kind of music library you will need. It is best for you, your clients, and the DJ industry if you only perform at the types of events that you enjoy, and for which you are qualified. As your business develops and you bring on other DJs whose talents differ from your own, you can expand the types of functions performed by your company.

Clubs and Bars

Bars and clubs present great mid-week or weekend booking opportunities for both club and mobile disc jockeys. But only if you can deal with the owners or managers who are all too often under-educated and unappreciative of our profession. The pay usually ranges from $100 to $200 per four- or five-hour shift, and the job frequently requires that a DJ bring her or his own equipment.

The DJ at a club or bar is central to attracting customers who want to dance. As long as the venue is making money, the owner is usually satisfied. The job at clubs requires extensive beat-mixing skills within a specific genre of music. At bars, music programming generally includes several genres, and a DJ's ability to promote drink specials, and sometimes even run special contests, helps please the client.

Locate a bar or club where you would like to work or where you would like to place your staff DJs. Make an appointment to speak with the owner, and sell him or her on why adding a DJ or replacing the current one with you will be good for business.

Negotiate the terms and conditions under which you will provide your services. Try to secure a six-month or one-year deal by offering a discounted price if the owner agrees to the terms. Be sure that the bar or club owner, manager, and you are all in agreement about the format and music that will be used and played.

Make certain to have a written contract that is signed by the owner, not the manager. And, if there are partners involved, each of them must sign the agreement.

Your contract should include the following:

- **Terms:** The length of the contract.
- **Payment:** How much and when you are paid.
- **Insurance liability:** The bar or club is responsible for all liability and property insurance, as well as music license fees.
- **Failure-to-appear provision:** If you are late or don't show up, the club or bar has the right to deduct a proportionate amount or all of your pay for that date.
- **Cancellation:** Your notification from the venue if it is closing due to inclement weather and to the venue if you cannot appear because of it.
- **Advertising:** The name of your DJ company and the DJ is to appear in conjunction with any advertising done by the venue.

- **Legal provisions:** Standard legalese that provides for the venue owner to pay legal and court costs, and interest in the event that you have to take the owner to court to receive payment, or the owner breaches your contract in some manner.

You should always seek advice from a legal professional who can design or look over your business contracts, to make sure they are legally correct and binding in your area.

Bar and Bat Mitzvahs

A mitzvah is a religious ceremony in which a Jewish boy or girl formally accepts responsibility for the commandments of Jewish law and adulthood.

It is not necessary to be Jewish to entertain at mitzvahs; however, it is essential to be highly familiar with Jewish customs, traditions, and word pronunciations. For example, the Hebrew word for blessing is "motzi," and you'd never want to refer to a "yarmulke" as a beanie.

This can become even more complicated when entertaining at events where the family is from another country. In these cases, you may have to learn how to do introductions in another language, such as Russian, Polish, Spanish, etc.

If you plan to work with the mitzvah market, be prepared to spend significant time in the planning stages of these affairs. In addition, a great deal of physical exertion is required during the games at the events themselves, so you'd better be in shape for them.

It is not uncommon to deal with clients whose children have attended others' mitzvahs, and want something different, bigger, or better for their child's celebration. Ask a lot of questions, and take copious notes during your initial meeting. Find out about the likes and dislikes of the parents and youngster. This can go a long way toward helping you provide entertainment solutions for a prospect.

Some mitzvah celebrations are highly religious in nature. Others are primarily parties. And there are clients who will want non-stop, high-energy entertainment, whereas others will prefer a more toned-down event.

Successful marketing of your entertainment services is especially important to capitalize on the lucrative mitzvah market, because the income potential is so high. Even for a basic bar or bat mitzvah, the revenue is significantly greater than with other types of parties and events. Prices can range from several hundred to several thousand dollars depending on lighting, staging, the number of DJs and dancers provided, and other add-ons such as big-screen video, karaoke, magicians, face painters, caricaturists, carnival games, prizes, and so forth.

Most people don't shop for mitzvah DJs out of a phone book. A marketing-savvy mitzvah mobile must learn the locations of synagogues within his or her area and whenever possible and establish rapport with rabbis.

Getting into this community generally requires advertising in Jewish newspapers and periodicals as well as in temple and Hebrew school publications. Check to see if they print a special bar or bat mitzvah issue.

Obtain membership lists from your local synagogues. Call or direct mail the families, to determine the age of the children. Two or three years prior to their mitzvahs, try to meet with the parents and mitzvah celebrant to sell them on your entertainment services.

Once you're firmly established in your local mitzvah market, marketing becomes easier and less costly because of the high percentage of word-of-mouth referrals from parents. Generally it is a tightly knit community that is not easy to break into.

To accomplish your goal, consider offering your services for free during Hanukkah or doing a one- or two-hour free party at a temple event where pre-mitzvah-age children are present.

Other ways to market your mitzvah entertainment services include doing party showcases, and low-cost or free "mock mitzvahs" at summer camps.

Corporations

Selling your services to corporations is an attractive proposition because the rate you can charge is usually higher than with private clients, and corporations offer the possibility of significant repeat business. Businesses usually book three months to one year in advance.

The corporate market is always in need of new, fun, interactive entertainment that will add to an overall event, and if sold properly, companies are willing to spend the money to get it. If there is no set budget, then you must educate the corporate contact about the services you provide. This includes the distinctions between professional entertainers and amateurs, and between those who are members of DJ associations and those who are not.

Corporate mailing lists can be purchased through a number of resources, including your local chamber of commerce, the Jaycees, convention and visitor's bureaus, convention centers, and the local chapter of the International Special Events Society. Advertising in business-oriented newspapers, direct mail, and telemarketing are other ways to reach business prospects.

Generally speaking, the human resources department of a corporation is almost always the first place to look. This department is in charge of things such as company holiday parties, employee appreciation events, training sessions, and other in-house events. Sometimes there is a committee in charge of planning corporate events. So be prepared to make at least one, two or even three presentations to close the deal.

Sometimes the job of hiring entertainment for events is given to someone as an extra responsibility. Few people consider this a top priority, because they are usually quite busy and often overburdened. So unsolicited letters, e-mails, and phone calls often go ignored.

The person in charge of making the arrangements wants his or her superiors to look favorably on the decision made. And the decision must usually be justified to someone else in the organization. Therefore, you need to provide ample evidence as to why you are a company's best choice as a disc jockey entertainer.

Whenever possible, have a concrete idea of a company's budget before submitting a proposal. Your proposal should include a low, medium, and high-end package. Think "outside the box" to envision the many different benefits and activities you can "bring to the table." These include a state-of-the-art sound system with wireless microphone(s), a huge music library, trivia contests, dance contests, team-building exercises (which are especially attractive to human resource department directors), raffle-prize giveaways, participation dances, and so forth. To be effective, your presentation needs to focus on the value of your services—not the price. It should emphasize both the tangible and intangible benefits both the company and the event attendees will receive as a result of hiring your services.

The typical corporate client will hire a DJ entertainer for a holiday party or company picnic. However, Fortune 100 and 500 companies often host major events and hire professional event planners, entertainment agencies, or destination management companies that provide a full range of conference and convention services to handle them. Get on their lists as a preferred DJ entertainer.

Sometimes your DJ services may be just a small portion of an extremely large budget that also includes a band, celebrity appearances, several cocktail parties over the course of a weekend, amusements, inflatables, and more.

Add-ons are a great way to pump up your profits and the energy level at corporate events. They enable you to tie in the entertainment with the theme, motivate the guests to participate in group activities, and create an exciting environment. Logo-imprinted hats, shirts, coffee mugs, and other items are just the beginning of some popular giveaways that can prove to be profitable upsells.

Hotels, resorts, and convention sites are also booking events for their corporate clients year-round. Picnics, award banquets, dinner dances, incentive meetings, and holiday parties are but a few of the parties needing entertainment.

In America, local car dealerships often have special sales for year-end blowouts and Presidents' Day. This is a great opportunity to sell your music and emcee services, as well as upselling prizes and giveaways.

Large charities sometimes host big events for their contributors and volunteers. Volunteering your DJ services or offering them at a reduced cost can be a great service to people in need and an opportunity to meet potential new customers who are attendees.

Schools

Entertaining at school functions presents an incredible opportunity to impress future mitzvah celebrants, brides, and grooms. By making these events wonderful and memorable, years later you could receive phone calls from people who are ready and enthusiastic about booking your services for their parties and weddings.

School gigs are physically and emotionally demanding because they require a lot of gear, energy, and patience. A youth-oriented DJ needs to be high-energy, interactive, willing to take lots of requests, and thoroughly knowledgeable about the latest hit songs that are likely to be requested.

Disc jockeys who entertain at school functions have to walk a fine line of keeping the students happy with the music they want, and the chaperones comfortable by not playing songs with offensive lyrics. If this is a niche market that appeals to you, be prepared to bring extra speakers and subwoofers, and a nightclub-style light show to school events.

Most school dances are held on Friday evenings in the gym or cafeteria. Proms are generally held at hotels, banquet rooms in upscale restaurants, and country clubs. Here is a listing of typical school occasions for which you can market your DJ services:

- Homecoming
- Winter formal
- Spring fling
- Junior and senior proms
- Graduation
- Dances
- Carnivals
- Picnics

Your fabulous wedding-oriented brochure just isn't going to cut it with the youth crowd. "Hip" is the key word when targeting this market. So, your marketing materials should be contemporary, colorful, and full of glitz.

Include color photographs that show you performing, your awesome sound system, and your outrageous light show. If a picture is worth a thousand words, then a video is worth a thousand pictures. Teenagers are impressed by large speakers with subwoofers and nightclub-style lighting effects. By all means use testimonials in your brochure. The copy should always refer to "students," never "kids" or "children."

You can obtain a listing of schools in your area (including addresses and phone numbers) from your local library, state department of education, or *Yellow Pages* directory. You can also conduct a search on the Internet.

It is best to begin your marketing to schools during the first week of school and then again right after winter break. Postcards sent quarterly will keep your name in front of the decision makers. Other than the student activities director, the contact names at schools change yearly. So be sure your mailing list is up-to-date.

Make a list of the elementary, middle, junior, and senior high schools where you want to entertain, then send your mailings to the:

- Student activities director
- Class adviser
- Student council
- Dance committee
- Presidents of the senior, junior, sophomore, and freshman classes
- Junior and senior prom committee
- Graduation night committee
- Parent and teacher groups

For colleges and universities, send your mailings to the:

- Fraternity and sorority activity chairs
- Minority groups
- Alumni organizations
- Campus organization activity chairs
- Office of student activities
- Office of residential life or campus housing
- Dorm presidents
- Student government entertainment chair

Additional information on college activities in the United States can be found through the National Association for Campus Activities (www.naca.org).

Weddings

Wedding receptions offer you the opportunity to make people's dreams come true and give them happy memories that last a lifetime (or a nightmare they will never forget!) That is why entertaining at weddings requires a great deal of performance and organizational skills, as well as the ability to cope with pressure and to act professionally in what can be the most trying of circumstances.

Note that renewal-of-vow ceremonies, holy unions, and same-sex wedding ceremonies are all included in the wedding category.

To be conducted properly, this type of event also requires top-quality equipment that is well maintained and reliable, a wireless microphone for toasts and blessings, and a music library with great variety that will appeal to every age group.

For these reasons a DJ should only pursue the wedding market when he or she has had a great deal of experience and is truly prepared with the proper tools. By joining a local pro disc jockey association, you will have the advantage of being able to gain feedback from peers about your readiness for these events.

Unfortunately, hiring a DJ entertainer is usually far down on the pecking order of vendor selections. Through consumer education, hopefully, one day this trend will change.

During a consultation and presentation, focus on the benefits you will bring to the prospect, and do not turn the meeting into an "It's all about me" spiel. It takes practice to develop good skills in this regard, but they will pay off handsomely in your sales closing rate and in the fee you can collect.

Sales meetings with wedding prospects often need to be conducted in person. This is because there is so much information to discuss with the bride and groom before they can come to an informed decision about which DJ to hire.

A bride and groom must be confident of your ability to provide them with an incredible entertainment value for their "once-in-a-lifetime" event. When you receive a telephone inquiry from this type of prospect, use every amount of persuasion you can to get them to agree to meet with you face-to-face.

If a bride and groom feel you are professional and competent, they may hire your DJ service. If you want to earn the really "big bucks" in the wedding market, here are the Top 10 things you need to do:

1 Use benefit-oriented language that appeals to the prospects' emotions and that assumes you will get the booking.
2 Ask questions to learn what the couple wants and doesn't want.
3 Listen closely to their responses, and take careful notes.

4 Conduct an organized presentation that verbally explains your services in detail, as well as showing them in print. Includes testimonials as well as the vendor coordination activities you will perform. Show a DVD of your past wedding performances that clearly demonstrates your abilities as an incredible entertainer and emcee and that people love to dance to the music you play.

5 Answer any and all questions, and overcome objections.

6 Verbalize your willingness to take the time, and place your energy into customizing their wedding reception according to their vision of it.

7 Provide ideas and advice for making their special day spectacular and memorable.

8 Be perceived as an expert who is a polished professional, and who can be trusted to provide exceptional entertainment, detailed event-planning services, and total peace-of-mind.

9 Have your contract available for signing, and ASK FOR THE SALE!

10 Deliver more than you have promised.

Charities

Doing charitable work is an excellent way to make new contacts and good karma for yourself.

My personal "rule of thumb" in whether or not I will entertain at a charity, is that the organization must spend no more than 10 cents for every $1 raised and no more than 10 percent of its operating expenses on fundraising. A good source to investigate this kind of information is www.charitynavigator.org.

You may choose to do one or two free events yearly and charge one third to one half of your regular corporate rate, depending on the expenses and time involved, for others. Be sure that in return, your company's name is included in all advertising, marketing, and public relations for the event. Always use a contract, and then use any press received in your marketing materials.

Be careful not to do so many freebies that it cuts into your profit margin. After all, there are only about a hundred Friday and Saturday nights a year.

It is recommended that American DJs not "trade checks" with charities, because when you get paid by them you are taxed as income at whatever tax bracket you are in. In addition, you can only deduct 50 percent of your donation. Disc jockeys in other countries should check on the tax implications in their area.

Karaoke

From ancient times in Japan, a party livened up whenever someone started singing. The other party-goers would keep time with hand-clapping, and it never mattered whether the person sang well or out of tune. In the case of the latter, it sparked laughter and made the gathering more lively.

Having such a custom, the Japanese are extraordinarily gracious when they listen to other people sing. They are generally not shy at all about singing in front of others. This is probably the main reason that karaoke has been largely accepted in Japanese society.

Karaoke is an abbreviated, compound Japanese word: "kara-" comes from *karappo*, meaning empty, and "-oke" is the abbreviation of *okesutura*, or orchestra.

It is widely believed that the use of karaoke started at a snack bar in Kobe City, Japan, more than twenty years ago. People say that when a strolling guitarist could not come to perform at the bar due to illness, the bar owner played tapes of songs, and encouraged the patrons to sing along. The innovation was such a hit that bar after bar picked up the contagious pastime.

Karaoke became a firmly established form of entertainment in Japan. Typically, businesspeople stop by a bar with colleagues after work, have a drink, and enjoy singing popular songs to the accompaniment of karaoke.

Karaoke stimulated people's desire to sing and has grown to be a nationwide amusement, thanks to technological development and a new business called the "karaoke box," which consists of compartments made by partitioning and soundproofing rooms in a building. Today, karaoke boxes are widely popular among all sectors of the Japanese population. Karaoke also plays a role as a place for family communication through singing.

Originally in the form of tape of a popular song's accompaniment, karaoke has grown to be a major entertainment industry through technological innovations such as the video disk, laser disk, and CD graphics.

The karaoke boom has spread abroad, enjoyed not only in Japan but also in Korea, China, Southeast Asia, Europe, and the United States. Because karaoke displays the words and scenes of a song on a monitor, it has also been attracting the attention of countries trying to improve their literacy rate, as a good educational tool.

Listening to music, watching TV, and using home computers have made people become passive receivers of entertainment. The advent of karaoke (and growing popularity of mobile DJ entertainers) may well help correct this trend and make a great contribution to the history of musical entertainment.

The karaoke business operates on a very basic premise: Everybody wants to be a star. You don't have to be a singer to be a karaoke DJ, but it certainly helps if the entertainer can "get the ball rolling."

Karaoke equipment and supplies are now easily accessible to consumers through major retailers. The good news for karaoke DJs is that we can offer "entertainment extras" that the typical person cannot. For example, it is very difficult to create a "live stage feeling" at home. We can bring extra TV sets, lights, stage props, and even costumes to enhance performances.

For DJs who do not have day jobs, karaoke presents an excellent opportunity to secure weekday gigs.

Karaoke singers generally tend to follow their favorite DJ from venue to venue. However, many of these folks do not care for alcohol or bar and club environments. A karaoke show can bring excitement and added entertainment value to a multitude of venues and arenas, including:

- Bars and nightclubs
- Private clubs
- Organizations and clubs
- Shopping malls
- Car dealerships
- House parties
- County fairs
- Coffee shops
- Retirement communities
- Schools

- Weddings
- Roller skating and ice skating rinks

The easiest way to market and sell your karaoke services is through an effective verbal and printed presentation, combined with a DVD or CD-ROM that shows the highlights of your and the patrons' performances at the type of venue where you are trying to get a booking. Hosting a contest for prizes or a coveted grand prize is a great way to attract patrons to any public karaoke gig.

A key selling point to venues where alcohol is served is that having a karaoke night will increase the number of patrons who come in, thus adding to liquor sales. Retail locations can benefit from increased traffic to their location, which can add to the stores' sales. With private events and retirement homes, the primary focus of your sales pitch should revolve around the increased entertainment value karaoke will bring because the attendees enjoy it so much. Evening karaoke jobs can last for several hours, whereas day events rarely go longer than one or two hours.

Bridal Fairs

If weddings are a mainstay of your business, then becoming a vendor at bridal fairs in your area is very important to the marketing efforts of your business.

Bridal shows are usually an expensive investment. So before signing a contract, find out how many other DJ companies will be at a bridal fair. If there are more than a couple, it may be wise to pass on a particular bridal show.

Bridal shows generally attract a highly targeted group of women who are interested in expanding their knowledge about wedding-related services and products. Such events present an excellent opportunity for you to personally speak with many potential clients, which is a great advantage in sales.

It is not enough just to be present. You must direct traffic to your booth through large, attractive signage that is designed like an advertisement to make the benefit stand out. Your use of lighting, promotional items, and drawings for free prizes will all help attract people to your location.

If you are a solo operator, work the booth with a significant other or complimentary vendor. If you are a multisystem operator, persuade a couple of your DJs to "work the booth" with you on a rotating basis, dressed in formal attire. You can entice your DJs by letting them know that this is a great opportunity for them to personally get booked for a wedding.

Here are some ideas for maximizing your bridal fair sales opportunities:

- Offer a $100 to $150 discount to anyone who books your services within two weeks of the show.
- Continuously show your company wedding DVD or CD-ROM
- Conduct a drawing for a free DJ. Have visitors complete a slip of paper that includes their name, the name of their fiancé or fiancée, address, telephone numbers, e-mail address, and wedding date. This drawing will give you an excellent mailing list to market your services to after the bridal show.
- Make arrangements for onstage showcase opportunities to conduct mock wedding party introductions using your sound and light system.
- Everyone working the booth should be aware not to put their hands in their pockets or cross their arms.
- Use elegant, gold-tone photo frames to present your marketing materials.
- If space allows, have a sound and light system set up at your booth.
- If you are allowed to play music, use instrumental cocktail or dinner tunes.
- Lay down an area rug inside your booth to add a touch of class.
- Don't place a table between yourself and your prospects.
- Offer the free use of one of your sound systems and DJs for announcements that need to be made. Do this in exchange for recognition of your donation from the show's producer, over the sound system.
- After each conversation with a prospect, give away a promotional item as a token of your appreciation.
- Use spotlights in your booth to draw attention to your sound system, literature holder, and free drawing location.

In addition to direct sales opportunities, bridal shows are also great places to network with other wedding vendors. Make sure you have enough helpers for coverage at your own booth, then try

DJ Wes Flint stands ready to greet prospects at a bridal fair. Note his beautifully displayed marketing materials. Photo courtesy Wesley L. Flint, DJ Wes' Mobile DJ Service

to meet the other exhibitors and collect their business cards. A word of caution: NEVER interrupt someone who is speaking with a prospect.

Sound Reinforcement

For most DJs, weekdays are "dead" times when their sound and lighting gear lies idly in a storage area. You can change this by adding sound reinforcement as an additional revenue stream for your business. Many folks just need a basic public address system, which a mobile DJ can easily provide—plus a whole lot more, if needed.

There is a huge amount of money up for grabs in the sound reinforcement and equipment rental market because of the millions of dollars corporations spend annually on events where they need this gear. Let's face it, the house sound systems at hotels usually leave much to be desired.

There is a better alternative. They could be hiring your DJ service to meet their needs at a lower cost and with greater service than most equipment rental companies' offer, so buck the trend!

Most local sound rental companies charge a set fee for a basic sound system, and charge extra for additional items such as wireless hand-held and lavaliere microphones, etc. Check out their fees, and consider making yours 10 to 20 percent lower to break into the sound reinforcement market.

To obtain business, do direct mailings to corporations in your area. Contact banquet facilities and hotels, and send them a sound reinforcement rate sheet (along with your other marketing materials). Meet with the person in charge whenever possible.

Here are your three primary sales points:

- You (or a DJ from your company) will personally monitor the sound or lighting at the event.
- Your sound reinforcement equipment rental rates are 10 to 20 percent lower than the rates of other companies in your area.
- The quality of your gear is equal to or better than what others are offering in the marketplace.

*If you work just for
money, you'll never make
it. But if you love what
you are doing, and always
put the customer first,
success will be yours.*

RAY KROC

6

Customer Service: A Contact Sport

As a DJ, you need to sincerely be in tune with your client's emotions at an event. You are not there to "clock in, then collect your check." Clients appreciate a DJ who really cares, and who is invested in providing exceptional customer service.

"PARADISE" MIKE ALEXANDER

Photo courtesy
Mike Alexander,
Paradise
Entertainment

Owner: Paradise Entertainment Inc., Orcutt, California
Speaker: DJ Times International DJ Expo

Velcroing Your Client

Have you ever walked up to a salesperson and then have to wait for his or her attention for no apparent reason? Have you ever gone through a maze of automated options with your long-distance provider, trying desperately to speak to a "live" person? Though the financial benefits of providing great customer service are widely known, some companies—especially large one—seem to blatantly ignore the facts.

Every time you greet a prospective customer, remember that there is plenty of competition in your business for their business. One of the best distinguishers you can supply from the first greeting onward is the sense that your clients *matter*. Let them know this about you, and you will have gone 70 percent of the way to making a sale and keeping a client for years to come.

If you're looking to distinguish your disc jockey service by offering something that will be a real competitive advantage, then focus on customer service. If you can establish a reputation for great service, it will be easier for you to get new clients, get more business from your existing ones, and raise your prices.

Rule #1 in providing superior customer service is to give people more than they expect. Go ahead, promise the moon and the stars—but only if you can truly provide them. If you give only lip service and do not follow through on your commitments, you are still providing customer service—the kind that will send clients running to someone else for their next event.

Giving clients *more* than they anticipate will not only gain you repeat bookings but it will also enhance your reputation and lead to referrals. When clients are impressed, they tell their friends, colleagues, and acquaintances. When they are *not* impressed, they tell anyone who will listen. On average, it costs a small business ten times as much to attract a new customer as it does to retain an existing one.

When someone hires your DJ service, they automatically expect reasonable rates, a complete music selection, a professional performance, and pleasant service. Go the extra mile. Promise less and deliver more, then watch the clients line up at your door.

One of the greatest thrills in business is acquiring a new client. It's easy to get so caught up in the excitement of acquiring new clients that you do not spend enough time or money on unlocking the value of your existing ones. After you have been in business for a while, responding to the needs of your existing clients must take priority over prospecting for new clients.

Acquiring customers is important, but retaining customers is critical to the ongoing success of your company. Whenever you have a business totally dependent on new clients, you're vulnerable. If economic conditions change or a new competitor enters the market, you may suddenly see your customers disappear. So, what types of things can you do to create a renewable income stream?

Follow-up, follow-up, follow-up. This is the key to retaining customers while growing your business. Like anything worthwhile, consistent follow-up requires a lot of effort, but over time you'll reap the benefits of a steady stream of repeat business and referrals. Studies show that it takes far less time and money to sell to an existing client than a cold prospect.

Customer service is a "feeling" the clients experience, not a thing they can see, touch, or smell. They know it when they feel it. You never want to lose a satisfied customer, so look for ways to keep and serve them on a repeat basis. Your bank account will thank you.

Generating Referrals

Referrals are the best way to grow your disc jockey business because the referring party has presold your entertainment services. Ultimately, you may have to close the deal, but the prospect-to-client conversion rate is much higher with referrals than cold calls.

Many DJs find that after a couple of years, the most significant percentage of their business comes from referrals. The most successful disc jockeys have a formal program to try to increase their referral business.

A happy client who tells others about your service is your best advertisement, so you want to foster this. Past clients are in a direct position to tell everyone they know about the great services you provide.

Here are some prime opportunities for contacting past clients after events:

- Immediately after the event, send a thank-you card or letter along with a Client Satisfaction Survey with a postage-paid envelope, a transferable $25 or $50 coupon that can be used at their next event (or a friend's), a couple of business cards, and a brochure.
- For wedding, anniversary, birthday, and mitzvah clients, send them a birthday or anniversary card with a personal note.

STACY ZEMON
ENTERTAINMENT
"Bringing the *Party*
to Your Event!"

CLIENT SATISFACTION SURVEY

Dear Valued Client,

Your answers to the following questions will help us learn what we are doing well, how we may improve, and if your expectations were met. As a token of our appreciation for completing this survey, we will send you a coupon worth $50 off the cost of the next event you book with us. And, we will extend this same courtesy to any of your referrals who engage our services. After completing this survey, please send it back to us in the stamped envelope provided. THANK YOU!

Stacy Zemon

CLIENT NAME: _____   **ORGANIZATION:** _____

TYPE OF EVENT: _____   **EVENT DATE:** _____

WHAT PROMPTED YOU TO CALL STACY ZEMON ENTERTAINMENT?

_____ Past Experience as a Client

_____ Saw Performance at an Event (Whose, Where & When?) _____

_____ Referral (Whose?) _____

_____ Brochure (Where did you obtain it?) _____

_____ Advertisement (Where & When?) _____

Other _____

IF AN AD PROMPTED YOU TO CALL, PLEASE RATE ON A SCALE OF 1 TO 5 (1 = MOST), EACH REASON THAT APPLIES.

_____ Immediately Grabbed My Attention   _____ Motivated Me to Act Quickly

_____ Explained Benefits of Choosing Your DJ Service   _____ The Image Drew My Eyes to Your Logo

Other _____

WHAT PROMPTED YOU TO CHOOSE STACY ZEMON ENTERTAINMENT? PLEASE CHECK EACH REASON THAT APPLIES.

_____ Past Experience as a Client   _____ Liability and Property Insurances

_____ Saw Performance at an Event   _____ American Disc Jockey Association Membership

_____ Referral from a Trusted Source   _____ Availability of Payment Via Credit Card

_____ Professional Experience & Attitude   _____ Emergency Backup Equipment & Personnel

_____ Event Consultation Services   _____ Author of Books for the DJ Industry

_____ Availability of Light Show, Party Props, etc.   _____ Value of Services for Rate Charged

_____ Female DJ Entertainer

Other _____

HOW MANY OTHER DJS DID YOU SPEAK WITH BEFORE BOOKING STACY ZEMON ENTERTAINMENT?

_____ Before Calling Stacy Zemon Entertainment   _____ After Calling Stacy Zemon Entertainment

- CONTINUED ON OTHER SIDE -

RATE THE FOLLOWING:
(NOTE: Your ratings should reflect the feedback you received from the MAJORITY of your event's attendees!)

	EXCELLENT	VERY GOOD	GOOD	POOR	N/A
PERFORMANCE:					
Entertainer Personality	_____	_____	_____	_____	_____
Emcee Skills	_____	_____	_____	_____	_____
Music Selection	_____	_____	_____	_____	_____
Attire	_____	_____	_____	_____	_____
TECHNICAL:					
Sound System Audio Quality	_____	_____	_____	_____	_____
Light Show Quality	_____	_____	_____	_____	_____
Equipment Appearance	_____	_____	_____	_____	_____
CUSTOMER SERVICE:					
Attitude	_____	_____	_____	_____	_____
Helpfulness	_____	_____	_____	_____	_____
Promptness	_____	_____	_____	_____	_____
Time Spent	_____	_____	_____	_____	_____
ENTERTAINMENT EXTRAS:					
Light Show Value	_____	_____	_____	_____	_____
Streamer Launcher Value	_____	_____	_____	_____	_____
Party Props Value	_____	_____	_____	_____	_____
Karaoke Value	_____	_____	_____	_____	_____
Big Screen Video Value	_____	_____	_____	_____	_____

PLACE A CHECK NEXT TO THE CATEGORY THAT MOST CLOSELY MATCHES YOUR OVERALL SATISFACTION WITH STACY ZEMON ENTERTAINMENT

_____ Extremely Satisfied   _____ Unsatisfied
_____ Very Satisfied   _____ Very Unsatisfied
_____ Somewhat Satisfied

DESCRIBE THE POSITIVE ASPECTS OF YOUR EXPERIENCE THAT STAND OUT, OR KINDLY MAKE A RECOMMENDATION FOR IMPROVING OUR SERVICES

_____

_____

_____

_____

MAY WE USE YOUR POSITIVE COMMENTS FROM ABOVE IN MARKETING OUR SERVICES? _____ Yes   _____ No

WOULD YOU RECOMMEND STACY ZEMON ENTERTAINMENT TO OTHERS? _____ Yes   _____ No

IF YOUR ANSWER TO THE ABOVE QUESTION IS "YES," KINDLY PROVIDE REFERRALS … AND THANK YOU!

NAME   TEL. NO.   BEST TIME TO CALL

_____

_____

- New Year's cards can be sent to any client, and do not risk offending a client who is not Christian by sending a Christmas card.

Depending on the success of your referral program, you may want to include incentives such as promotional items, movie tickets, or gift certificates. Such items are highly memorable to most people and can go a long way toward having past clients become ambassadors for your disc jockey service.

Fee-based referrals are a perfectly legitimate aspect of doing business. Viewed as profit sharing, it is simply compensating someone a fair fee in return for a referral. Deciding whether to have a formal or informal relationship depends on the frequency of the referrals, the size of the jobs, and the agreement between the two parties.

It is not at all uncommon for a mobile entertainer to pay a 10 to 20 percent fee or "kickback" to banquet facility managers, photographers, videographer, or other professionals in return for their referrals.

Prior to making a referral, some banquet managers and caterers require a disc jockey to carry liability and property insurance. If you want to encourage these folks to refer you, then it is suggested that you request a tour of their facilities, meet with them personally, offer the company a financial incentive, and above all, be easy to work with.

For instance, check with facilities to make sure they allow fog machines, hazers, and/or confetti cannons before attempting to upsell these items to a client. In many cases you will need to educate a caterer or banquet manager about the differences between a fog machine and hazer. In addition, I recommend that you use streamers, not confetti, because it is far easier to clean up.

On arrival at an event, greet the banquet manager or caterer and hand him or her an itinerary that has been approved by the client. Few DJs do this, and it is a sign of your professionalism and organization that will likely be appreciated. The itinerary also ensures that all critical event personnel are "on the same page."

Last, but certainly not least, be consistent about your referral fee policy. For example, offering it to one limousine service and not to another could damage your credibility. Some banquet facilities expressly forbid their managers to receive kickbacks under penalty of losing their jobs, so be careful when approaching the issue. Ask directly if it is appropriate and acceptable to them.

Keeping in Touch

There are many positive ways to provide customer service besides "putting out fires." An excellent method for creating "top of mind" awareness of your DJ service with past clients and vendors is to produce a newsletter or e-zine on a quarterly basis. Websites for DJs and party planners are filled with articles of value that you can use to fill out the pages.

Most authors will grant you permission to use their articles free of charge in your newsletter or e-zine as long as you include a link to their website or credit them as the author. Include a personal letter to your customers, information about your latest products and services, and testimonials. Pictures from events, coupons, expert tips, etc., all fill the pages of your newsletter or e-zine with opportunities for your client to interact with you again. Sending newsletters or e-zines with valuable tips and information provides a great opportunity to remind your clients that you value their business.

Another important tool to keep in touch with past clients is sending notes, along with two business cards, every six months to a year. You can include appearances you are making in public venues as an invitation to see you again and have some fun.

Mail a Client Satisfaction Survey a week after every gig. Make it easy for the client to return it to you by enclosing a self-addressed, stamped envelope (SASE). Most people will be happy to oblige your request, and the survey will provide useful information about the client's experience with you or one of your DJs.

In the survey make sure to ask for permission to use the client's name and comments in your marketing materials, but also make it clear you want the comments whether quotable or not. As you build up your "raving fan" collection, share it with prospects.

Internal Customer Service

Are your employees loyal ambassadors of your DJ service? In many ways, your employees should be considered high-ranking diplomats representing your company. They must display the same diplomacy you would when a disagreement surfaces, and they can act as gracious hosts when greeting your clients, just as you would if you were there.

Empowering employees to handle problems that happen at events benefits your business in several ways. Research has shown that employees who have this kind of freedom begin to think more strategically about their work and about your business. They endear themselves to your clients because they act as client advocates. They go beyond satisfying needs, to exceed expectations. This leads to happy and repeat customers, which gives you a competitive advantage.

The primary reasons people stop doing business with a company are because they weren't treated up to their level of expectation, or they were treated with indifference. Never let this happen just because you or your employee had a bad day. Consider holding monthly staff meetings. (Providing pizza is a strong motivator for people to arrive in a timely fashion.) Create a forum in which staff feel free to share information with you and their co-workers. Use these meetings to constantly reinforce the benefits to them and to the business of superior customer care.

Good communication is key to the success of any kind of relationship. The more employees know about the goals of your DJ business and how they can personally contribute to accomplishing those goals, the more they will feel like part of the team. Reward positive input and on-the-job performance with compliments, lunch, bonuses, gift certificates, and other forms of appreciation that are important to the person being rewarded.

Be certain that your passion for customer service runs rampant throughout your company. Employees should see how providing excellent service relates to business profits and to their futures with the company.

Handling Complaints

More business is lost due to simple misunderstandings than any other single reason. No matter how thorough you are and how much you follow through on planning, inevitably complaints will occur. Many of these issues arise just out of the process of having to coordinate so many diverse elements in planning, and through the lack of follow-up by others, even when yours is perfect. So, what's the best way to handle client complaints? Welcome them.

When people are angry, upset, or frustrated, they often act in an emotional manner. Emotions block rational thinking. It may be difficult for clients to vocalize their complaints effectively when

Top 15 Things That Stacy Zemon Will <u>Never</u> Say or Do at Your Event!

1. Charge a cover and check for ID.

2. Use a Super Soaker to point out people on the dance floor.

3. Conduct bridal party introductions in Pig Latin.

4. Table dance while dinner is being served.

5. "I interrupt this slow song for a fabulous puppet show."

6. Secretly fill the room with laughing gas and fog.

7. Emcee operatically.

8. Release a flock of doves inside the banquet hall.

9. Pass a collection basket and tell people that giving is good for their souls.

10. "If you want me to play your song request, you'll have to sing it to me."

11. Incite a revolt in the name of the guest of honor.

12. Talk about politics and religion between each song.

13. Play "Duck, duck, duck, duck... GOOSE!" with the guests while they're dancing.

14. Wear clown makeup, a clown wig, clown shoes, a clown nose, and *nothing* else.

15. Weather reports every 10 minutes.

DJ Stacy Zemon has a strict code of behavior when performing at events.

they are being led by emotion rather than logic. In addition, some people are just critical, demanding, or obnoxious.

Although any of these scenarios may test your patience, it is important to wear a smile and maintain a thoroughly professional attitude. Whoever first said, "The customer is always right," knew a great deal about customer service. The first response to any complaint should be to *listen* without interruption, *validate* the person's feelings, then *commit* to promptly attending to the situation and *remedying* it (if even remotely reasonable). Staying positive with negative clients will make it easy for them to give you honest feedback.

Develop a culture in your company that treats every complaint as the key to developing a better way of doing things. As a DJ business owner, it is extremely important to let all your staff members know how strongly you feel about customer service and to let them know you will hold them accountable for delivering the best.

Here are some simple rules for handling complaints:

- Listen attentively to what your clients say and how they say it.
- Respond courteously and directly.
- Identify the basis and cause of any complaint.
- Give specific feedback about what steps you will take to respond, including the research you will perform, the vendors or services with whom you will discuss the difficulty, when you will get back to the client, and the method you will use for contact.
- Set a reasonable timeframe for the resolution.

Here, for example, is a typical complaint situation you might encounter. Your client, "Gigi," says that the banquet manager is refusing to let you introduce the wedding party. You offer to personally call the banquet manager and to make her or him aware of the client's wishes and of your experience in making VIP introductions. Promise Gigi that you will call her back to give her the results of the conversation. Then ask her, "Are you comfortable with this solution?" With luck, she will reply that she looks forward to hearing from you and receiving the confirmation, as it takes the argument away from her to resolve. She will also indicate if Tuesday is a good timeframe, or she will make an alternate suggestion. Either way, you know she understands what follow-up you will pursue.

Of course, many resolutions may require research between the event suppliers before a response can be formulated. After researching the matter and discussing it with the appropriate parties, if you determine that the complaint is valid you might also think about avoiding such complications in the future by making constructive changes to your own event-planning guidelines.

To effectively implement constructive changes, details about the problem need to be discussed with the right people on your staff.

For instance, in the example just given, contact with the banquet manager that includes a wedding reception itinerary agreed to by your mutual client would go a long way to avoiding the banquet manager's objections.

In the case of a problem that cannot be met with an immediate solution, here are some phrases that can help you win back an unhappy client until the problem is solved:

- "I can appreciate what you're saying."
- "I can understand how you'd feel that way."
- "I can see how that would upset you."
- "It sounds as if we've caused you inconvenience."
- "My understanding of what you've just shared is . . . "

When you or one of your staff members has not performed up to the client's expectations, be sincere in your concern, apologize for the specific behaviors or actions that have upset the client, and provide a resolution that is acceptable to the client. Even if the problem has been generated by other vendors, apologize for the inconvenience to the client and, when possible, assure the client you will make the problem disappear.

Let the client know that you will personally take responsibility for the situation. Ask, "What would make this situation right for you?" In most cases your good listening skills, responsiveness, and apology will be sufficient.

When it is not sufficient, make sure the client knows what you are going to do to remedy the situation. Explain the actions and timelines you need to take to make things right. Be certain to respond in the timeline you have specified. If you promised follow-up by the end of the day, get back to the customer before the end of the day, even if you have no solution yet. The update will still lower the client's anxiety about a solution to come. As a rule of thumb, double your estimate of the time it would normally take to get the information before offering to report, to ensure that you resolve most difficulties before the first call-back.

When seeking a win-win situation, get the client's agreement by asking,

- "Would this be agreeable for you?"
- "Is this the solution you were looking for?"
- "Will this make things right for you?"

After the situation is positively resolved, be sure to ask the client to do business with you again. You have earned the right to ask this question because you demonstrated that you are a caring, sincere, proactive professional. Why *wouldn't* someone do business with you again?

Closure of a customer's complaint is just as important as solving the problem. Done correctly, it creates a positive resolution that converts someone from an unhappy client to a repeat customer. Unfortunately, some clients cannot be pleased, no matter how hard you try. Be as cordial as possible; however, if you reach a point of diminishing returns, it may be time to move on and focus on more profitable clients or prospects. Some valid reasons for reaching this point may be that the client:

- Doesn't respect or appreciate your efforts
- Makes excessive demands on your company and staff
- Is not fair-minded in his or her expectations
- Tries to take advantage at every turn
- Views you as a disposable vendor and not as a valued entertainer
- Wants a full refund based on a minor problem

Even in these circumstances, never lock the door behind you when you go. There's no point in telling a client, "You're more trouble than you're worth." Even a bad client may refrain from causing you undue bad word of mouth if you treat him or her with respect when you walk away. Try to walk away with a smile and on friendly terms.

Refund Policy

Consider instituting a "100% satisfaction guaranteed or your money back" refund policy. Such a policy can help you convert prospects into clients because it conveys your confidence in the quality of your service. You may want to add a caveat that your policy is subject to the terms of agreement in your contract.

If you find yourself refunding more than 5% of your sales to dissatisfied clients, then you have a service or performance problem that needs your immediate attention. In this case, it is worth every penny it costs you to identify and fix the problem before it has a more major impact on your disc jockey business.

You can turn this loss into a gain by learning from the experience. Take every client complaint very seriously. Thank the clients who let you know that they're unhappy. Try to find out what happened and how you can make it better. Asking the client is always the right decision. Remember, any refund policy that depends on the legal system is a loser for you and your clients. The big winners in the courtroom are usually the attorneys.

*If you're trying to
persuade people to do
something, or buy
something, it seems to
me you should use their
language, the language
in which they think.*

DAVID OGILVY

7

Advertising: Made You Look!

To create an effective advertising campaign you must
research and choose mediums that will effectively
reach your target audience(s). The copy must be
concrete, easily understandable and contain a specific
"call to action." Over time, combining these elements
will yield quantifiable results/dividends and your
business will flourish.

SCOTT KILEY

Photo courtesy
Scott Kiley

Publisher: PRO SL Directory—The Yellow Pages for Pro Sound, Lighting,
DJ and Event Production

Advertising Essentials

The purpose of advertising is to increase the exposure of your business. It should also motivate qualified potential buyers to call and inquire about your disc jockey services. A well-planned and properly executed advertising program should include a sufficient commitment of capital resources. It is an investment in future profits.

Advertising does more than directly impact immediate sales; it also reinforces the buying process, which can take several weeks to months. By carefully selecting the most beneficial media, you can promote your service to thousands of potential new clients annually. A prospect has to experience your ad often enough to remember it, then have a need to hire a DJ before contacting you.

Advertising can be one of the fastest ways to market and grow your business, or it can be one of the quickest ways to go out of business. With the right ads you can attract clients to your business and increase your profits. With the wrong ads you can spend your way into bankruptcy.

To grow your business you need to attract the attention of your prospects. Advertising can help you do so if used correctly. Unfortunately, many DJ business owners waste thousands of dollars on advertising efforts that only achieve minimal results.

If you want to get the most from the money you spend to promote your entertainment services, make sure to have a strong message that speaks to your prospects' needs and wants. Use each ad to identify at least one common problem of your prospects and the benefit of using your DJ company. Before placing your ad, test several versions out among your network of friends and family, to determine which one gets the best response.

Your advertisements should be eye-catching, look professional, and convey the most important aspects of your disc jockey service. These aspects can be service, price, quality, or special entertainment options that you offer.

If your message fails to stand out, even the best-placed ads will fail to get attention. Keep your messages consistent across your ads, and stick to one or two main selling points. Your ads must motivate prospects to act *now*.

For example, if you want to prompt prospects to visit your website, include an offer that entices them to do so. Of course, even if you are successful in attracting people to your site, this doesn't guarantee that you will make sales. Make sure the content and visual organization encourage visitors to take the action you want them to (i.e., book your services online).

When to Advertise

Start your advertising campaign three to four months before the season or type of events for which you want bookings. For corporate holiday parties, begin advertising in January of that year, as corporate budgets and parties are planned at least a year in advance. For wedding receptions, begin advertising in mid-December, when many couples announce their engagements. Advertise again in March to attract their attention for summer wedding receptions. For spring dances for schools and colleges, you begin advertising in early fall.

A well-coordinated effort could keep you busy throughout the year with a wide variety of events. Of course, fraternity parties and other private celebrations occur year-round, so keep ads geared toward these markets running continually as long as they are producing results.

Copywriting

Great copywriting comes from knowing who your prospects are, what they need and want, and how your service will specifically give it to them. Like any piece of written material, your ad should have a title, a beginning, a middle, and an end. To write an effective ad, you must know your objectives and the target of your ad.

A proven formula for effective advertising is "Attention, Interest, Desire, and Action." This proven plan (1) gets the prospect's *Attention*, (2) fosters his or her *Interest* in your offer, (3) builds *Desire* for your service and (4) generates some type of *Action* on the part of the buyer.

Attention (the headline): Keep the headline simple and short, with strong, attention-getting words that convey a powerful benefit statement. This and your graphic are the most important elements of your ad. You have only a moment to capture a reader's interest, if you want that person to read further. A powerful headline will (1) stop the reader (2) isolate and qualify your best prospects, and (3) pull your reader into any subheaders and body copy.

How do you write an attention-getting headline? First, carefully review all the benefits of your service. Second, take your most important benefit and weave that benefit into your headline. Use action words to describe the benefit to one individual reader. Find examples of powerful headlines that do the trick and appeal to you in your own reading, and use them as examples of how to frame your own. One caution—avoid being overly cute or offensive. It may capture a reader's attention, but it will not get you that reader's business.

Interest and desire (the offer, body copy, benefits): Be specific and include as many benefits as possible in simple and interesting terms, however, be very careful not to overwrite. People do not want to read columns of information to find out the key points they need to know. You can use phrases, bullets, and graphics to pull the reader into your message without a lot of text.

Identify with the customer's need, and be clear about how you can fulfill it. Make a compelling offer such as a coupon or other incentive. Offer free party-planning information, and describe why it is so useful.

Words generate interest and graphic design displays the words in a visually appealing way. No beautiful design will make a poorly written ad sell for you, but good design reinforces good copy. Copy and design should work synergistically. If you are unsure whether your ad is pleasing, brief, clear, and attractive, try asking your friends or business acquaintances whether they like it and get the point.

Action (ask for the order): Advertise a special offer for a limited time only to motivate prospects to act immediately. You can make it easy for them to do so by offering secure online booking. Take fear out of the purchase by giving solid guarantees and providing testimonials from satisfied clients. Remember to always include your phone number and website address.

Word-of-Mouth

When was the last time you told a friend about a good movie? You may not have realized it at the time, but you were providing the best possible advertising for the movie studio: word-of-mouth advertising. No other type of advertising is more powerful than word of mouth. Your friend may have dismissed all the typical Hollywood hype, but if *you* say it's a good movie he or she will probably go see it.

Of course, word-of-mouth is a great booster for a good movie, but the opposite is also true. If it's a bad movie, the negative word will spread even faster than the positive, and people will stay away in droves. This is doubly true for a service business.

For every three people willing to spread the positive word, there are thirty-three others who will spread the negative, if you give them cause. The best way to build on the business that your advertising dollars bring in, is to offer exactly the service you advertise, and to do so with expertise and warmth. No amount of advertising can replace the "goodwill" you build by providing excellent services at a reasonable price.

If you want clients to spread the word about your disc jockey service, then all their experiences with your business must be positive. Your performance, customer service, and follow-up must all be first rate.

In the business world, positive word-of-mouth generates referrals and referrals lead to sales. So word-of-mouth praise is an important marketing tool that can contribute to your success. You love to be the recipient of positive word-of-mouth and so does everyone else. So when you come across people doing their job well, be sure to "sing their praises" both to them and to others.

No matter how great you, your employees, and your DJ business are, you cannot please everybody. So don't be surprised by the occasional complaint. Be ready to fix whatever the problem is immediately and pleasantly. Even if fixing a problem costs you money, in the end it is an investment in your future, because a disgruntled customer can cost you plenty of business with poor word-of-mouth remarks. Did you ever go back to a place that refused to fix your problem? Probably not. A customer whose problem has been fully handled becomes a satisfied and happy customer. That customer will walk away smiling and become an ambassador for you and your company.

Here are some ways to avoid problem situations and please most of your customers most of the time:

Be True to Your Word: You must be trustworthy and do what you promise to do for your clients. Do not make false claims or promises. Your clients will remain loyal, if you are true to them.

Build Relationships with Your Clients: Get to know your clients as people. Talk to them as you would to a respected friend. But keep it sincere. Do not put on the big "Slap on the back, nice to see you" act. People are smart enough to distinguish real enthusiasm for your work from phony sales pitches.

Appreciate Your Clients: Have contests, give away freebies, do whatever you can to make sure your clients know that you appreciate them. A simple "Thank You" can do wonders, and a special gift or unexpected discount can do even more.

Provide Help for Your Prospective Clients: If your clients need advice or answers, make sure you do your best to see they get that help. You need your clients more than they need you.

Treat People with Respect: Treat others the way you would want to be treated. A special word of caution here—whatever your ethnic or cultural background is, you will be dealing with people from diverse and different backgrounds from your own. Be careful about inappropriate humor or potentially offensive political, religious, sexual, or ethnic references. Many a customer has been lost by the wrong joke at the wrong time, even when well meant.

Providing exceptional performances coupled with "beyond the call of duty" client service, can have a positive effect on your business. However, *not* providing what your clients want can bring down the curtain faster than you can blink. Successful word-of-mouth advertising means satisfied clients telling your story to others. If they are treated fairly and courteously, they will.

The least expensive form of advertising is word-of-mouth, because it costs you absolutely nothing. Keep in touch with your former clients. If they just got married, chances are many of their friends are getting married, too. Also, beef up your referrals by networking with other vendors: caterers, photographers, etc. After every event, collect business cards and then send your network a "It was nice working with you" note, with a stack of your own business cards. This professional word-of-mouth follow-up can also be a real business booster.

Yellow Pages

This is an expensive, but sometimes necessary, form of advertising. Most telephone books have specific sections headed "Disc Jockeys," "Entertainers," "Entertainment Bureaus," "Special Event Coordinators" and "Wedding Consultants."

An ad in your local *Yellow Pages* is necessary because it lends credibility to your business, assuring the public that your company is not a fly-by-night outfit. In my opinion, however, it doesn't take a large display ad to establish that fact—a listing to a medium-sized ad is just fine.

Display advertising is most highly recommended for multisystem/multilocation operations, and for DJs who are seeking a greater quantity of clients by offering lower prices in a marketplace.

Most *Yellow Pages* shoppers are seeking the lowest price, so although they may assume the company with the biggest ad is the best, they may also assume this company is the most expensive. Some statistics show that most people who use the *Yellow Pages* will call between five and ten companies before making a decision.

To maximize the effectiveness of your ad, work with the salesperson to get the best placement for it. The upper right-hand corner of the page is the most sought-after spot, as it is the first to be seen while turning pages. The upper left-hand corner is the second best placement. Statistics show that consumers look first at the largest ads, ads with color, and companies whose names start at the beginning of the alphabet.

If a display ad is not in your budget, consider a bold business listing that includes a super slogan, a fabulous logo, and the use of color or anything that makes your service stand apart from the competition.

Some disc jockey services have only a one-line business listing, which is free with a business telephone line. Their reasoning is that they feel people who use the *Yellow Pages* are only shopping for the best price, and such people aren't their demographic target.

A one-line listing makes you accessible to people who are seeking you out and who do not have any of your marketing or advertising materials. Sometimes former event attendees will recall your name or the name of your company, but they don't know how to contact you. The *Yellow Pages* are likely to be the first place they turn to find you.

Here are some important factors to consider when advertising in the *Yellow Pages:*

- How many leads per month does it generate?
- How much do you pay monthly for the ad?
- How many of your competitors are running classified ads?
- What size are the ads that your competitors are running?
- Are you listing in major metropolitan area or in small county books?

The fewer the disc jockeys who are advertising, the greater the benefit of advertising is to you. For cross-promotional purposes, be sure to include your website address in your ad. Although you can also advertise in the *Yellow Pages* on the Internet, most DJs do not find this a cost-effective form of advertising.

Radio

Radio advertising is only recommended for multisystem and/or multilocation operations. The exception to this is in cases where you are bartering your DJ services for air time. Radio stations often do remote broadcasts that require a professional sound system. You may be able to work out a trade deal by providing the system in exchange for commercial production and air time. If you buy or barter for airtime, do so only on stations with formats that are most listened to by your target audience(s).

A well-produced commercial can include mood music. It can also include testimonial "snippets" from satisfied clients. Consider having a company jingle produced for use in all your radio advertising. Sell the benefits of your unique DJ service. Tell listeners about your experienced DJ entertainers, state-of-the-art equipment, and gigantic music selection. If you offer party- or event-planning services, dancers, video and lighting effects, or karaoke, mention this in the commercial as well. If your service is geared toward specific types of functions (weddings, bar or bat mitzvahs, corporate events, etc.), mention these as your areas of specialty in the radio "spots." You may want to develop more than one ad for use on different stations with different market profiles. Ask stations to provide their demographic information, as well as numbers of audience at various times of day or night. Be sure you are reaching the audience you want to promote.

A 10-second commercial is useful to keep your company's name familiar to the audience. It can comfortably accommodate twenty to twenty-five words. A 30-second commercial permits some explanation of a sales idea and at least one repetition of your business name. It can accommodate fifty-five to seventy-five words. Don't forget to get your telephone number or website into the ad more than once.

Television

Purchasing television commercials is much the same as purchasing radio commercials. American cable TV advertising is far less costly than advertising on the major networks. Although your ads will reach a smaller audience, cable television allows for placement on specific programs and can be targeted to defined geographic areas. Wedding programs that are viewed by brides-to-be, or MTV, which is viewed by potential school clients, may be appropriate channels for placement for your commercials.

Television account executives have software tools available that can assist in determining precisely what TV shows, stations, and time slots will best allow you to reach your target audience(s).

Many television stations will write, produce, "shoot," and edit a commercial for you free of charge or at a low charge, if you advertise with them on an ongoing basis. This can be beneficial if you are working on a tight budget. If you choose this option, make sure you have approval rights on the completed commercial in your contract. Also be aware that the production quality will likely not be as good as if you hire a professional production company to shoot your commercial.

Retaining the services of a professional videographer or production company is costly, but the product may well be worth the cost. Today's viewing audiences are sophisticated and expect to be visually "wowed." Using graphics, creative visual images, and music will gain their attention.

Be certain to not include anything in your TV commercial that will date the production or is a variable that could change. And never quote rates or fees in a commercial. Once you have gone to the expense of producing the spot, you may want to run it repeatedly over time. If this is the case, have your commercial produced as a "donut ad." A donut ad has an unchanging opening and closing, and a changeable center or "hole." The hole can be filled with information that changes from season to season or year to year. It can also be filled with information geared toward certain markets or events.

Another production technique is to leave a few seconds at the end for a specific event promotional message or to plug an appearance. This can then be added in whenever you want it included. For instance, you could add the fact that you are appearing every Saturday at the Reflections Night Club. The time can also be used to link your ad with someone else's event, such as "DJ Jerry is a proud sponsor of the Atlantic Golf Classic, coming February 10th to the Whitemarsh Golf Club."

Whichever style of ad you pick, remember to keep the message simple and clear, and make it easy for anyone listening to get in touch with you.

Publications

Confine your newspaper advertising to affordable weekly papers, where you can run your ad by postal code, covering only the geographic areas you want or can afford. You can also run ads in the neighborhood sections of local daily newspapers, but they are generally more expensive than weekly papers.

Since large-circulation newspapers are usually read more completely by the over-50 segment of our population than by the younger folks, you may want to confine your general-public newspaper advertising to specialized wedding sections or education supplements only.

Here is a listing of advertising sources through which you can target various audience(s):

- Newspaper bridal guides
- Local wedding guides
- Religious publications
- School newspapers
- Newsletters published by special interest groups
- Weekly local newspapers, community shoppers, and penny savers (under "Entertainment" in the classified section)
- *Parents Express* and other parent-oriented publications

Direct Mail

Direct-mail options include single-piece mailings or the more affordable, cooperative mailings from companies such as Val-Pak in the United States. In a single-piece mailing, your letter arrives by itself. The pieces mailed in the cooperative envelopes are all the same size, so no one company looks larger or more established than any of the others.

A single direct-mail piece stands alone but is more expensive to produce. The cooperative piece is crowded into an envelope with numerous other pieces, and the recipient must shuffle through the others to get to yours, but it's a lot less expensive. Match your choice to your budget.

Your single piece can be mailed whenever you wish and as often as you please; the cooperative piece will be mailed in compliance with a structured mailing calendar. Val-Pak is the leader in direct marketing. It offers advertisers marketing strategy and developing, ad concept, graphic design, printing, and distribution. Online offers are available in conjunction with mailings. The packagers can vary versions of your offers by zip (postal) code for optimum results and tracking.

To use this medium effectively, you may want to offer a $50 or $100 discount coupon, free light show, or extra hour of perform-ance time at an event. Be sure to make your offer for a limited time only, to encourage the prospect to act quickly.

Boxes, tubes, and just about anything that isn't flat grab are called "dimensional mailers." They attract the attention of recipients, who virtually always open them. Think about how much fun it is to open presents at holiday time, and it's easy to understand why boxed mailings to business-to-business prospects are so successful!

Dimensional mailers capture and keep the attention of your prospect for longer than other "flat" mail. They can help you stand out from your competition. Some marketers report response rates of 25 to 50 percent or even higher, which is extraordinary. (Typical direct mail receives an 8 to 10 percent response, at best.)

Probably the biggest challenge is deciding what to put inside your box. You can use a promotional product that plays music or flashes lights when opened or even make it a "party package," complete with confetti and other accoutrements.

One large drawback to dimensional mail is its per unit cost. However, if you have a small target group of top prospects that represent significant sales potential, the high per unit cost is largely irrelevant. The key is to reserve your dimensional mailings for well-qualified lists. Call to make sure you've pinpointed the correct recipients for your mailing. It is recommended that you focus your dimensional mail on corporate, wedding, and bar/bat mitzvah prospects.

Here are some tips for making your dimensional mail campaign a success:

- Use your company colors, name, and logo on the package.
- Clearly label your dimensional mailer. A poorly labeled or "mysterious-looking" package may be discarded because of security concerns.
- Include a small gift such as a pen, a magnet, etc., that the recipient will keep to remind them to call you.
- Follow up by phone three to five business days after you have sent a mailer.
- Include a call to action that has recipients do something, such as visit your website for an additional gift or extra bonus.

Be aware that mail that doesn't meet U.S. Postal Service size guidelines is subject to a service charge. Contact your local postal center to find out what its guidelines are.

Here's a helpful hint. The nearly one hundred Postal Business Centers around the United States will perform a one-time free service: they will take your mailing list (up to 50,000 names) and insert the correct postal bar codes and standardize addresses.

Websites

Advertising your DJ service online can be a great asset for attracting new clients. Websites offer the flexibility of allowing you to have several pages of information available for potential clients to look at 24/7. Unlike two-dimensional advertising, you can even include short video clips of your performances. You may want to establish two or more sites that are targeted at different clients, such as a site geared toward corporate clients, wedding clients, and school clients. This will allow you to better customize services for each audience.

Many companies have high-speed Internet access for their employees. For convenience, prospects often go online during their workday to do research for their event. Some may not be able to call you from their workplace, which is why it's a good idea to have a form on your website for visitors to fill out so you can contact them at home. Always indicate the hours you may be reached at your phone number, as well. Always have an answering machine or service during off-hours.

Another benefit to having a Web presence is that prospects planning an out-of-town engagement will likely not have access to that area's local *Yellow Pages*. Although they may think of using www.superpages.com, they are equally likely to type something like "Philadelphia wedding DJ" into a search engine.

Also consider whether to have a generic host or a host who specializes in website hosting for disc jockeys. Hosting from a company not specializing in DJs can cost as little as $10 monthly. This is generally no-frills hosting that includes only a few e-mail addresses and about 100 MB of space. If you choose a company that specifically hosts DJ websites, the cost is higher, but so are the returns. Industry-specific hosts can often help with the promotion of your site and provide you with an interactive tool for your clients. The cost averages about $25 to $30 monthly.

Several websites are specifically designed to help DJs gain exposure. They simultaneously assist party planners to find entertainers and to book them quickly and easily.

Statistics indicate that customers need to see an online advertisement at least seven times before they will buy. Keep the same ad, or make sure they all have a common theme or image, to keep a memorable identity. If you constantly change your ads for no valid reason, you may be losing your identity with potential customers.

Internet-based professional listings and banner ads of various types can provide a great source of referrals for your business. Some websites offer free listings with paid upgrade options. Others are free of charge, and you only pay them a commission if they are successful in booking your events. Here is a listing of the most popular sites that the general public can visit to hire a DJ:

- www.800dj.com
- www.directcatering.com
- www.discjockeys.com
- www.discjockeyonline.com
- www.dj.net
- www.djapproved.com
- www.djyp.com
- www.findyourdj.com
- www.freedjamerica.com
- www.gigmasters.com
- www.mobiledjnetwork.com
- www.PartyPros.com
- www.ProDJ.com
- www.ProDJfinder.com
- www.TheDJlist.com
- www.usodja.com

There is only one thing in the world worse than being talked about, and that is not being talked about.

OSCAR WILDE

8

Public Relations: The Holy Grail of Freebies

Left to right: Ryan Burger, Jessica Agard, Bryan Foley, Ken Heath, and Paul Beardmore Photo courtsey www.ProDJ.com

Volunteer your services for charitable events. Donate equipment that you no longer use to nonprofit groups. Help out and participate in community organizations. By taking these actions, your company can become known as the local DJ service of choice. Mine has! The primary cost of PR is your time, and it is an investment that can bring in tons of new clients for your business!

RYAN BURGER

Owner: BC Productions, Des Moines, Iowa
Founder: ProDJ.com
Member: American Disc Jockey Association

Objectives

Public relations (PR) is the most effective way to build brand name recognition. PR is, literally, the relationship your disc jockey company has with the public. This not only includes prospects, but also future clients, employees, peers, vendors, other businesses, and the press.

The relationships you create with these people will have a profound effect on the future of your business. The function of PR is to publicize your company to these folks in a positive manner, which will translate into increased sales.

One of the objectives of public relations is to become known as an expert. What better way to be deemed an expert than to be published or mentioned in a prominent local—or better yet, a national publication or a book? Which do you think your prospects would rather choose? One of twenty local DJs they found in the phone book, or someone who's been quoted, published, or appeared in a well-known publication as an expert?

If you are quoted in a print or electronic medium that is noteworthy, you can cite "as seen in (or on) . . ." in your advertising, marketing materials, website, and e-mail signatures. "As seen in *Mobile Beat* magazine" or *"The Wall Street Journal"* can give you tremendous credibility and set you apart from your competition in a big way.

Being quoted or featured in the press gives an air of celebrity status for a local disc jockey company. Once again, this truly separates you from your competition.

It is important to remember that public relations is not marketing, although it can play a key role in marketing your DJ services.

Avenues

Public relations is much more than just a press release. A planned approach that uses many different methods will increase your probability for success and generate the kind of PR you want for your disc jockey business.

There are several PR avenues that can increase the top-of-mind awareness among your current clients and target prospects.

As always, this is most effective when used in conjunction with other advertising and marketing strategies.

Chief among the opportunities to gain visibility for your DJ company are the following.

Speaking Engagements: Speak for free to audiences that are part of your target market. This can include Rotary clubs, chambers of commerce, and trade associations. Public speaking engagements are like making a sales call to many people at one time. They give you instant credibility as an expert, and people like to buy from experts. Many times the speaking engagement is publicized ahead of time, and sometimes the media will show up at such events. Avoid speaking in July, August, and December, as attendance is generally low during these months. You can cut fixed costs by doing joint-venture marketing. Understand that the majority of your money will be made on the "back end"—by booking events.

How-To or Advice Articles: There are loads of weekly and daily newspapers, business magazines, trade publications, and print and electronic newsletters to which you can pitch your topics of interest. Articles don't have to be long, they just need to be informative. Whenever possible, maintain the copyright so you can offer the same articles to different publications. If writing is not your forté, there are plenty of freelancers who can ghostwrite articles for you under your name.

Newsletters: Writing a newsletter is another way to keep your name in mind. This can be online (e-zines) or printed. Newsletters contain content valuable to your target market and information about your products and services. Send them to your client and networking databases through e-mail or regular mail.

Television and Radio: Do you think TV or radio is too expensive as an advertising medium? Not if you are being interviewed or calling in on a talk show. Then it's absolutely free. Compile a list of answers to anticipated questions or FAQs for producers.

Free Publicity

Free publicity can be a valuable tool to raise visibility, boost sales, and help launch a new product or service. Your public relations efforts are more than a process for attracting new customers. Free publicity also:

- Builds top-of-mind awareness of your name for future clients
- Demonstrates the continued activity of your business
- Reminds existing customers of your services
- Establishes credibility

One important way to receive this publicity is by submitting newsworthy press releases to the media. Press releases that lead to free publicity can impact increased profits for your business. Seek out and create opportunities to appear in the media. This will enhance your credibility and the public's awareness of your mobile DJ service.

If you don't trust your talent for writing a good press release, it may be worth hiring an independent publicist. Not only can this type of pro create future ideas for your DJ business, but he or she will write the release and submit it to the media for you.

When one of the media does a story on your company, it's almost like getting a third-party endorsement. Articles, unlike advertisements, don't cost you a dime (unless you hire someone to do it.) Before you can send out a press release, develop your "angle," a good reason for writing the story.

You can create a newsworthy event rather than waiting for one to happen. For example, if you are a multisystem operator, consider forming a company sports team and playing against radio and television stations. By doing this, you will very likely gain additional exposure through the media with whom you are competing.

Coupled with an effective advertising campaign, being proactive with seeking out free publicity opportunities will fuel the sales cycle of your business.

Topics

A press release is the spark designed to prompt a reporter to want to write a story or an editor to assign a reporter to write it. Good press releases describe your services in terms of how people can

This ad is an excellent example of effective corporate public relations. It was placed on the outside back cover of the 2005 Mobile Beat DJ Show & Conference program guide. Courtesy, The American DJ Group of Companies.

Making It as a Mobile: Preparation and Inspiration

By Mike Buonaccorso

The Mobile DJ Handbook, Second Edition
by Stacy Zemon
(Focal Press)

In the Mobile DJ profession you can think of your performance as either a product or service. Either way, when what you offer your clients is consistently excellent, you will receive referrals that lead to increased bookings. Drawing on more than 25 years of combined mobile, club, karaoke, and radio experience, Stacy Zemon, one of the profession's most successful women, has created a handbook that can help you improve your performance and, in turn, your earnings.

Stacy Zemon

Now available in a revised second edition, *The Mobile DJ Handbook* emphasizes professionalism and contains useful information on buying equipment and music, securing bookings, and running party dances, contests and games. Zemon conveys this information in a concise, easy-to-read manner.

Those new to the DJ business, who lack the experience, references, and connections that benefit higher-priced, established mobile disc jockeys, can get up to speed quickly with this helpful guide. And one thing leads to another: learning tips to generate more bookings will create the cash flow that is necessary to promote, improve, and expand your business.

More than being a master of ceremonies, being a DJ requires at least a basic technical knowledge of the appropriate gear. An entire chapter is devoted to Mobile DJ equipment, which isn't that different than the equipment employed by club DJs—except that Mobile DJs also need portable power amps, speakers, and lighting equipment. In addition, building a comprehensive and diverse music library is essential. Zemon provides tips on this subject and provides some helpful lists of novelty, specialty, and participation songs. She also addresses bridal introductions, bar/bat mitzvahs, and other theme/holiday parties that require extensive preparation.

With chapters on marketing, advertising, customer service and more, *The Mobile DJ Handbook* is not only an excellent resource for aspiring disc jockeys who desire to be successful in this highly competitive profession, but also for those experienced pros who want to keep ahead of the pack.

The Mobile DJ Handbook, Second Edition is available online at www.focalpress.com or at a music or book dealer near you.

Performance Beyond Expectation
by Ray Martinez
(ProDJ Publishing)

At one time or another, most Mobile DJs who have been in business for an extended period of time have probably said to themselves, "I should compile all those notes I've been jotting down over the years and write a book based on what I've been through!" Well, one disc jockey veteran did just that. The result is *Performance Beyond Expectation*, by Ray Martinez of RayMar Productions in Anaheim, Ca.

Martinez, a disc jockey whose career has spanned over thirty years, shares personal stories, photos, anecdotes and his unique collection of mobile memorabilia in his book, which is published by ProDJ Publishing.

Starting in college radio in the early seventies, Martinez moved through a career encompassing many facets of entertainment. While much of the book reflects the author's strong spiritual convictions, it is, at the same time, a useful compilation of informative personal reflections and self-help tips

from others.

Performance Beyond Expectation provides an interesting collection of timeless material for long-time DJs who can relate to the author's observations on the unique nature of their chosen profession. It's also for those who are new to DJing who are looking for an idea of where the road might lead.

For information on ordering *Performance Beyond Expectation* contact ProDJ Publishing at www.prodjpublishing.com, sales@prodjpublishing.com or 800-257-7635.

PERFORMANCE BEYOND EXPECTATION

Personal stories, anecdotes and quotes to motivate DJs who want to improve their business.

Mobile Beat magazine review of *The Mobile DJ Handbook* and *Performance Beyond Expectation*. Courtesy, *Mobile Beat* magazine.

Author Stacy Zemon autographing a copy of the first edition of *The Mobile DJ Handbook* at DJ Times' 1998 International DJ Expo in Atlantic City, New Jersey.

benefit from them. Your services must offer solutions to their problems, answers to their questions, improvements to their lives, or savings on their time and/or money.

Think along the lines of newsworthy and interesting topics. A one-page press release that opens with who, what, when, where, and why will increase your probability of getting into a publication. Include some background information, a quote from you, and the contact information. That's all there is to a press release. If the contact wants more information, he or she will call or e-mail you.

If you're not sure about what you can write about in a press release, consider these ideas:

- Introduction of a new product or service
- Celebration of an anniversary
- Announcement that your DJ service is putting on a seminar that is free and open to the public
- Opening up a satellite office or location change
- Receiving an award

- Announcement of your availability to speak on entertainment and event-related subjects of interest to the general public
- Participation in a charity event
- Earning recognition in the DJ industry
- Establishment of a unique vendor agreement, partnership, or alliance
- Announcement that someone in your company has been named to serve in a leadership position in a community, professional, or charitable organization

Save all the press you receive to impress your clients. You can laminate and frame articles for display in your office, keep them in a photo album placed in your reception area, and by all means show them to prospects during in-person consultations.

There are other things you can do with press releases. You can post them on your website, and you can include them in direct-mail pieces to customers and prospects. Use your imagination and you will be surprised at the unique ways you have available to generate publicity and ultimately buzz about yourself and your DJ service.

Format

Here is the basic format for a press release:

- **Margins:** Type your release on $8\frac{1}{2}$- × 11-inch letterhead with 1-inch or $1\frac{1}{2}$-inch margins on all sides, using only one side of the paper.
- **Page numbering:** Do not number the first page, but number subsequent pages. Print the word "More" centered on the bottom of the page to alert reader there's another page to follow.
- **Pages:** Staple, don't paper-clip, the pages of your press release together.
- **Release date:** In the upper left-hand corner under your letter-head type "FOR IMMEDIATE RELEASE"; if it should be released on a certain date, type "FOR RELEASE" (followed by the month and the day). Use boldface type, capitalize every letter, and make this release date line a slightly larger font size.

- **Contact information:** Include a contact name, work phone, home phone, and fax number two lines underneath the release date.
- **Headline:** Leave two blank lines after the contact information, and list in boldface your main headline. Headlines should never be longer than a sentence.
- **Dateline:** The city or town your press release is issued from and the date you are mailing your release will begin your first sentence.
- **Lead paragraph:** The first paragraph needs to grab the readers' attention and quickly imparts the relevant information key to your message.
- **Text:** Indent five spaces (a tab space) for new paragraphs, or leave two blank lines between each paragraph and flush left. The main body of your press release should thoroughly develop your angle. Begin each sentence with action-oriented verbs. Limit paragraphs to two or three sentences. Use the present tense whenever possible, especially when you're describing specific service attributes. Include facts and quotes from key individuals. Do not carry over a paragraph to the following page.
- **Italics and bold:** Use italic and boldface to make key words stand out (but be careful not to overuse).
- **Ending:** At the end of the release, type three hash marks, # # #, centered on the bottom of the page. This lets the reader know your release has concluded.
- **Folding:** Don't fold your release like a letter; fold it in a Z shape so that the headline and date will be the first thing the editor or reporter sees on opening the envelope.
- **Photos:** If you have good photos, send them or include the words "Photos available on request" with your information at the top of the page. Only send high-quality photos, however, and only when they add to your story. Black and white 8- × 10-inch glossies are best, but 5- × 7-inch is okay too. Place photos on top of cardboard when mailing. Don't tape or paper-clip because doing so can ruin the photo.

Always stress the most important points first with the lesser points following. The first paragraph should answer the questions: *who, what, when, where,* and *why.* Proofread the release for spelling and grammar errors. Make sure your message is simple and clear enough to be understood by anyone.

See the "Sample Materials" section at the end of the book for press release examples.

Media Kits

A media kit, sometimes called a press kit, is an essential tool for any disc jockey business that wants to gain exposure in print, on television, radio, or the Internet. It is simply an information packet about your company. The goal of the press kit is to grab the readers' attention, make a lasting impression, and create enough interest that they will contact you for more information.

There are many items that can go into a media kit, depending on the situation, the audience, or the use. Only include information that is current and most relevant to your target reader. There is a growing trend toward online press kits, which can be made available on your website.

The basic elements of a media kit include the following:

- **A cover letter:** Make reference to the fact that the kit was requested or, if you haven't had any previous contact, state why readers should care about what you're telling them. Provide a table of contents or a brief description of the items enclosed in the actual press kit. Let the reader know you are available for follow-up interviews and questions when appropriate. Also make sure to include your contact information.
- **A fact sheet:** Provide a brief overview of you and your DJ service, in an easy-to-read format. It should describe your services, company history, key personnel, any other notable facts, and a contact person.
- **A press release:** This may be about any event of interest.
- **A list of services:** Include the types of events conducted by your company and information about how people can engage your DJ business's services.
- **A sample story:** Some editors will print a sample story verbatim, as they view ready-to-print articles as an easy way to fill up space with little effort on their part. They do, of course, usually edit these stories.

- **Frequently asked questions (FAQs):** The FAQs sheet helps the editor determine what questions to ask you in an interview or what to include in the article.
- **Photographs:** Include a professional photo of your DJs gathered together and another of a single DJ at a sound console at an event with people dancing.
- **Published articles:** What other media have done will be of interest to current media targets. This can include article reprints and printouts of online press that your company has received.

Photographs

It is important to have professional PR photographs to include with your press releases and media kits. Minimally these should include an above-the-shoulders photo and another "action shot" of you entertaining at an event. If you are a multisystem operator, also include a unique group shot that includes all your DJ entertainers. You will want color as well as black and white pictures.

JPEG (.jpg) and TIFF (.tiff) are both image file types—in the same way that an MP3 (.mp3) is a sound file type. JPEG is currently the most widely used image format for Web and multimedia use, whereas TIFF is currently regarded as the industry standard for digital print and press.

JPEG is popular on the Web because it uses a compression codec. It is actually a codec, not a code that keeps the file size small, which makes it easier/quicker to download.

Although small, heavily compressed, low-resolution images—most of which are around 72 dpi (dots per square inch)—may look good on screen, and be suitable for multimedia and Web use, they will not be of high-enough resolution for the types of digital and offset printing that are used in magazines and newspapers. For this, an image with a resolution closer to 300 dpi must be used.

The most common mistake made is that people presume because an image looks fine on their screen, that it will print the same way. If the resolution isn't fine enough, it won't: It will be pixilated, blurry, tiny, or all three. Once an image has been converted to low or screen resolution, it cannot be converted back to a high-resolution image—in the same way that a low-bit-rate MP3 cannot be converted back into a CD-quality file.

Stacy Zemon has some fun during a public relations photo shoot in the 1980s.

If possible, therefore, your digital images should be scanned as high resolution from the original source expressly for the purpose of print. From there, most commercial image-manipulation programs can convert and save them in a variety of formats and resolutions.

Most print and publishing houses will be happy to advise you on the formats they prefer to receive and the suitability of any digital images you may already have access to.

An image for printing should be delivered as high-resolution, 300 dpi (dots per inch) TIFF. If possible, scan it from the original source. Attempting to convert photographs that have already been converted for Web or multimedia use will generally not work very well.

If you look at a picture in a newspaper under a magnifying glass, you see it's made up of lots of small dots, and if you look at a picture in a high-quality glossy color magazine you see it's made up of many more, smaller, finer dots. Exactly the same thing applies in the digital world—except it's measured in pixels.

On the back of the hard-copy photos, place a label that includes the name of your DJ company, its address, phone number, and website address, as well as the names of the individuals appearing in the photograph. It is okay to include a suggested caption.

Some common uses for PR photos include your website, brochures, business cards, and other marketing materials.

Here are some tips to follow for a photo shoot:

- Have your hair cut at one to two weeks before the session.
- Style your hair and any makeup before you arrive at the studio so touchup is minimal.
- Avoid high-neck and sleeveless clothing.
- Wear solid colors or patterns that don't have small lines.
- Have a professional apply your makeup.
- Ask the photographer to use a solid, neutral backdrop. If you have dark hair, the backdrop should be lighter. If you have light hair, the backdrop should be darker.

Targeting Media

You need to have already identified your target prospects in order to distinguish the media that will best reach them.

Your local library has reference books in which you can identify the associations and organizations to which your potential prospects may belong.

Contact information for many national TV and radio shows can be found on the Internet. Thousands of media contacts also can be located in the Bacon's directories available in the reference section of

Denon & Doyle Disc Jockey Company.
Courtesy, *DJ Times* magazine.

most major libraries. There's one directory for each form of media: newspapers, magazines, TV/cable stations, and radio stations.

Media directories can become outdated quickly, however, so the best approach is to call each media outlet, tell them what the news release is about, and ask for the correct spelling of the name of the most appropriate contact.

Develop a database of contacts, using a popular software program such as M.S. Access or ACT, from magazines, newspapers, radio, and TV programs. It is important to include the names of editors and producers of each medium in your list. When submitting a press release, send it to every one of your contacts for which your news may be relevant. Be sure to know each of the medium's deadlines and the names of the people to whom you submit your release.

Many major media outlets accept electronic story submissions. Go to the show's website, and you'll usually find a link on the home page inviting you to submit an idea. Write a short e-mail, or send your news release in the body of the message—never as an attachment.

Follow-up

Getting your press release or media kit noticed is the first step to gaining publicity, so package your materials in an organized and professional manner. Commit to an ongoing effort to establish relationships with editors, reporters and producers. The more relationships you have with your targeted publications, the increased likelihood you have of getting publicity.

Each type of media has different requirements. Magazines, including trade press, have lead times of three to five months and are usually interested in advance features on noteworthy industry events and stories, whereas newspapers and radio stations are looking for local angles.

Send your press release to those in the database you have established. Some editors prefer faxed press releases, others prefer receiving them by e-mail. Learning about an editor's, reporter's, or producer's preference will increase your chances of getting noticed.

After sending materials, follow-up with a phone call to the recipient in three to five business days. Don't call at the end of the day. Editors and reporters are on deadline at this time of day and are scrambling to finalize their stories. Call in the morning, when they are likely to be more relaxed.

If you don't make contact, leave a voice-mail message. The key to success with editors and producers is to be persistent without being pushy. Don't leave too many messages. Voice mail is a good opportunity to leave a detailed message, but don't leave ten. It would be annoying to the journalist.

If you do make contact, use the opportunity to build relationships with your contact. Ask immediately if it's a good time for a quick conversation. If not, find out what a better time would be. If the contact will talk, promise to be brief.

Never start the conversation by asking, "Did you read my press release? I sent it to you last week." A good way to start a conversation would be "I noticed the article you did last month on the rising cost of hiring a band for weddings, and I thought your readers might be interested to learn there's an alternative that is not only less expensive but often preferred by event guests."

Being Interviewed

A well-placed quote or an on-air interview can position you as a sought-after expert in your field or even catapult your company to national prominence. It may take you months to pitch and win a prized interview—or you may be surprised one day by a call from an important journalist. And the way you handle your few minutes of fame can be crucial to your business's success.

When being interviewed by any form of media, it is extremely important to be prepared so that you come across in the most favorable light possible.

The chief motivation of most interviewers is to get information or create programming that's of special interest to their readers, viewers, or listeners. It's your job to provide relevant content while at the same time weaving in your own principal PR themes. No matter what you're asked during an interview, it should always be possible to lead the journalist back to a major point about your DJ service that you want to convey.

Here are a few tips to help you to prepare for these golden opportunities:

- When a reporter calls, take time to research the reason behind the interview request. Find out why that person wants to talk to you.
- Treat media representatives courteously. Being profiled in a news feature can be great for business. So be engaging and interesting, and stay focused on the purpose of the interview.
- Agree ahead of time on the length of the interview and the exact discussion topics. If your schedule changes, notify the reporter immediately. Breaking an appointment at the last minute won't exactly inspire the reporter to say wonderful things about you or your business.
- Understand the news media. Try to learn something about the various media that might contact you—print, audio,

video, and Web—and when the day comes that they contact you, be as open as possible. Avoid saying "no comment," and realize that with rare exceptions, *nothing* you say is "off the record."

- When you're being interviewed, give the journalist your undivided attention. State your messages clearly and concisely, and be able to support them through facts and figures, anecdotes, expert opinions, or examples.
- Control what you can. Be on the lookout for ploys designed to create controversy, and always bridge a negative point with a positive one.

For best results, rehearse and memorize the main points you wish to get across during an interview. When speaking them, your comments should sound spontaneous and thoughtful. Consider any interview a prime opportunity to favorably influence your company's image.

Think ahead of time about questions you are likely to be asked, and practice answering them. Have someone test you with mock interview questions until your delivery is smooth and conversational. If you have an on-camera interview coming up, it is helpful to videotape your rehearsal. Then watch the tape for any verbal responses or tics such as fidgeting that you will want to change.

By preparing your content well in advance and rehearsing your delivery, you'll also sidestep the anxiety or stage fright that can come from being ill prepared.

Promotional Items

Imagine giving the greatest performance of your life at a party and then having the guests say, "Great DJ, but I have no idea who that was!" Not a happy thought, right?

That's why it is important to use promotional items to gain exposure and publicity for your mobile DJ service. Consider putting your company's name and logo on your company vehicle, jackets, polo and T-shirts, refrigerator magnets, key tags, mugs, and so on.

In addition, have a tasteful banner with your company name and logo made professionally. Affix it to the front of your sound system or a banquet table using Velcro® or clips. The banner identifies your DJ service, serves as a decoration, and can hide wires and connections. Banners are appropriate at most occasions other than wedding receptions or formal affairs.

Providing promotional gifts to prospects and clients increases the likelihood of them booking your services or providing you with referrals.

*To effectively
communicate, we must
realize that we are all
different in the way we
perceive the world and use
this understanding as a
guide to our
communication with
others.*

ANTHONY ROBBINS

9

Networking: The Art of Tongue Fu

Photo courtesy
Bob Blanchard,
RJB Photography

Make a serious commitment to become involved in formal and informal networking activities. By giving and receiving information, support and camaraderie, you will benefit both personally and professionally. Actively encourage people to share their own thoughts and visions to a conversation. Remember to always be polite, friendly, tactful and open-minded in everything you say and do. To receive respect, we must first be willing to treat others in the same manner.

MATT PETERSON

Owner/DJ: Peterson Productions Disc Jockeys, Northampton, Massachusetts
Director: Professional Disc Jockeys of New England
Member: National Association of Mobile Entertainers

Effective Networking

One way to describe networking is as the establishment of a strategic alliance. This is collaboration that combines the expertise and/ or resources of individuals or organizations. There are several ways to collaborate that can bring added value and revenue to your business.

To network effectively, you must first put yourself in the right places. By getting involved with different organizations in your area, you will come in contact with many people. As you shake hands and talk with these folks, you will likely learn what they do for a living and where they work. Although the person with whom you are speaking may not personally hire the entertainment for the company picnic, she or he is likely to know who does.

Aside from informal, social networking, there are various types of organizations where more formal networking can occur:

- Business associations
- Religious organizations or clubs
- Special-interest groups
- Fraternal societies
- Charitable fundraisers
- Employee events
- DJ associations

Popular organizations include the chamber of commerce, Rotary Club, Lions Club, and Business Network International.

Many people join organizations for the sole purpose of seeking out referrals and sales opportunities. If this is your primary goal, it's quite obvious to the people you meet. There is a way to achieve prosperous results and help others achieve theirs as well. I call it "authentic personal marketing." Another way to look at it is "relationship marketing," which implies a long-term commitment rather than a quick "How do you do. Can you bring me business?"

Focus your networking efforts on meeting people with the purpose of learning about the quality products and services they or their companies provide. Why? So that you can recommend those products and services to your friends, family, acquaintances, and business associates.

This philosophy of generosity will be appreciated by everyone with whom you come into contact. People may express their appreciation by providing you with referrals and hiring you for their events. Even if they do not refer you, they will put in a good word when others bring you up in conversation. This type of goodwill is priceless for your business. It is my firm belief that if you want something for yourself, you must also give it to another. This includes always speaking well of people you meet through this type of networking, as well. If you give people a good reference or just a good word, the effort will come back to you many times over.

Another way to make rapid contacts within a business community is to consider being a greeter at the door for an organization's event. Volunteer your time for some of the group's activities. Join a visible committee and, of course, offer your services to emcee announcements at events. All these strategies can pay off in the number of contacts you make and the good impression people retain of you for future reference.

Here are some tips for making the most out of networking functions:

- Arrive looking your best.
- Wear your name tag on your shoulder, so people can read it as they hear you say your name.
- Develop a distinctive "signature" such as a snazzy tie, hat, or piece of jewelry.
- Develop an unforgettable greeting such as "I help people have fun for a living."
- Hand out and keep other people's business cards, and send a note and brochure to the people you have met.
- Travel with your own "PR agent" by attending with an employee or friend. This person can start the "good word" about you with others.
- Ask quality questions to start conversations. (Make sure to avoid controversial topics, if you do not know the direction the other person might lean.)
- Have something to say that is of interest and topical, but make it positive and light. Dreary subjects do not promote a feeling of well-being in other people about you.
- Show a sincere interest in the person with whom you are speaking.

A final piece of advice is to give qualified leads to other organizational members. Those that give the best leads to others usually receive the best leads in return. Remember, being memorable starts with repetition. Just like in advertising, repetition builds reputation. Happy networking!

Your USP

Highly effective salespeople have a USP (unique selling proposition). You need to be able to verbalize your USP in a clear and concise manner during any networking or sales opportunity. This description about who you are and what you do is your "elevator speech."

It is a 30- to 60-second sound bite that could succinctly and memorably introduce you during a short elevator trip. It spotlights your uniqueness. It focuses on the benefits you provide. It is authentic, enthusiastic, and invokes interest in the person with whom you are speaking. And it is delivered effortlessly.

A great elevator speech makes a lasting first impression, showcases your professionalism, and allows you to share briefly about who you are and what you do for a living. It can be used in formal business situations, in the supermarket checkout line, or when you're out buying your morning coffee.

To craft a great elevator speech, here are some action steps to take:

- Write down a brief explanation of your DJ services and your target market(s).
- List the benefits that clients derive from hiring you, and your competitive advantages.
- Use those phrases/words to create your benefit or outcome statement.
- Create an opening statement (hook) that will grab the listener's attention. It should be tantalizing, provocative, and even funny or a bit mysterious.
- Add a closing that is an open-ended question, which will invoke a response other than yes or no from someone. It should be a question whose answer provides meaningful information to you.
- Practice your finished elevator speech in front of a mirror. Record it and listen back.
- You will know you have succeeded in creating a good one if you come across as sincere, confident, and engaging.

Here's an example of an elevator speech that accentuates my USP:

Hi, I'm Stacy Zemon. I bring the party to people's events as a mobile DJ entertainer. By playing great music on professional sound equipment and acting as the emcee, I motivate people to dance and have fun. For the past 20 years or so I've successfully entertained tens of thousands of guests at private and corporate events as well as once-in-a-lifetime celebrations. People trust me because I wrote the world's best-selling book about being a professional DJ and am also a member of the American Disc Jockey Association. When was the last time you were at an event that had a DJ entertainer?

Depending on the value of the experience, you may want to give away a small promotional item such as a pen, along with your business card, to reinforce making the experience a memorable one. Be sure to ask for the other person's card, as well.

Complementary Businesses

A complementary business is one that offers products or services related to your business that may be of use to your clients, such as videographers, photographers, banquet managers, caterers, bridal stores, tuxedo shops, etc.

Having found some potential business types, choose ones that suit you in terms of:

- Proximity to your DJ service
- Amount of retail traffic or booking engagements
- Popularity or size of the business
- Its marketing presence
- Professionalism and attitude toward customers

Once you build a relationship with some complementary businesses, there are various ways you can help each other reach potential customers:

- Exchange links and banners on websites
- Share advertising costs (e.g., print ad, direct mail, expo exhibiting)

- Distribute marketing materials for each other at your places of business
- Swap articles for newsletters and websites
- Sell a product for and barter work from each other
- Review each other's products or services

By working with some complementary businesses, you can increase your potential client base for much less effort and money than most advertising methods.

VIP Contacts

Create a very important person (VIP) contact list comprised of key clients, referral sources, trusted advisers, and others who make your business and personal life run more smoothly and profitably. Be sure to include staff members, friends, relatives, and anyone who is a loyal ambassador for your disc jockey service.

Today's flexible contact management and database software packages make it easy to get personal with your VIPs. It is now a breeze to catalog detailed information, code your client lists, and sort them by virtually any criteria imaginable. Keep track of important personal dates, such as birthdays or anniversaries, and mark the events with a personal gesture. For convenience, stock up on greeting cards for various occasions, and keep them on file. Sending cards on nontraditional holidays such as Thanksgiving instead of the December holidays can make you stand out from the crowd.

If you read something that may be of interest to a VIP, clip it out and send it along with a personal note. When you learn that a VIP has won an industry award or made it into the press, acknowledge the good news. In addition to sending a note of congratulations, you may want to ask the VIP's permission to mention the news in your company newsletter or other venue to publicly acknowledge the event.

When a client or a colleague refers a new client, always send a note of thanks. If a VIP mentions getting over a cold or leaving work early for a child's soccer game, refer back to this the next time you speak with him or her. People are flattered when you inquire about their well-being.

Simply being thoughtful can build those all-important relationships that make or break your business. Investing a few moments in communicating personally with your key audience can turn contacts into contracts—without a hard sell. If you continually benefit from your contacts without paying attention to building relationships with them, you risk alienating this important group.

Barter

The value of a good trade deal cannot be underestimated. These deals can save you thousands of dollars of out-of-pocket expenses, if they are used wisely. Look in the *Yellow Pages* under "Barter Services" for barter houses that specialize in setting up such exchanges.

Barter services charge an initial sign-up fee ranging from $100 to $500. There is usually a smaller annual fee as well. Membership generally includes your DJ service's name printed in their directory and the ability to do trade deals with other businesses who are also members. Most barter exchanges allow you to buy an additional directory ad to stimulate interest in other companies trading with you. Some barter exchanges are networked with both local and national businesses.

Here's an example of how barter works. The owner of a local restaurant decides she'd like a DJ for her 25th anniversary party. Mrs. Restaurateur locates a disc jockey who is a member of her barter exchange. She pays the DJ's fee by transferring an appropriate number of "trade credits" to the DJ's account (typically $1 equals one trade credit.) Her trade credits have been accrued from trading goods or services with other barter members. She also pays a cash surcharge directly to the barter exchange (usually 10%) for their administrative handling. Any applicable sales tax is paid directly to the seller by the buyer.

Check how many other DJ companies are members of a barter exchange before joining. If there are more than a few, it may be wise to look elsewhere. You can also choose a barter exchange that has members who offer specific products and services you need.

Just be aware of one thing when you barter—taxes. In the United States, the government looks at barter as taxable income. If you barter more than $600 annually, your barter company will send you IRS Form 1099-B, which must be filed with your regular tax return. Even so, barter deals are a tax-deductible business expense in many cases, and they generally mean a large cash savings to you in terms of obtaining the products and services you need to carry on your business. Inquire about the barter taxation situation in your country.

For disc jockeys or DJ business owners, in addition to working with a barter company, strategic alliances often consist of simple "bartering" with corporate clients and suppliers for jobs or equipment and with competitors for overflow business. This simple bartering can often lead to a very valuable resource to keep your business stocked with the tools of the trade. Also consider doing trade deals with corporate clients, organizations, and associations. In some cases it is beneficial to offer your entertainment services at a reduced rate or at no charge in exchange for products, services, advertising, and memberships.

If you decide to use the services of a barter company, obtain references and thoroughly research its credibility. The investigation is necessary; otherwise you may find yourself in the position of doing jobs to accumulate points, with no way to cash them in for products or services you want. Before you barter, confirm that the business getting your trade credits is reputable and established. Most of all, make sure the list of businesses includes products or services you need.

Bartering with competitors can have its advantages as well. For example, say your expertise is in weddings, or you simply prefer to entertain adults. A call comes in from a prospect who wants to hire a DJ for a school function. In this scenario, you either subcontract or refer the job to a trusted competitor. When the "shoe is on the other foot," she or he does the same for you. This type of barter arrangement is advantageous to both parties.

A note of caution; do not use barter to an extreme. Cash flow is the life's blood of every business, and if you do too much trading you could be eliminating your opportunities to improve cash flow, which in the end hurts your disc jockey business.

DJ Associations

An ever-increasing number of DJs are joining professional disc jockey associations. This is an excellent trend, because it gives us opportunities to pool our ideas and efforts to improve the level of professionalism within our industry. I am a proud member of the ADJA.

Membership in a professional disc jockey association will also help you to stand out from the less professional competition in your marketplace—allowing you to convert more prospects into clients.

The American Disc Jockey Association and the National Association of Mobile Entertainers both have local chapters. On a more local level, regional and state associations often have meetings that include technical clinics, training, and open-forum discussions about ways to raise professionalism and combat "bottom-feeders" in the market. There is a friendly sharing of information, camaraderie, and even field trips and participation at charity events. Another benefit of joining some DJ associations is the ability to have access to equipment in the event of an emergency. A major benefit of most disc jockey associations is that they offer a service to members that provide referrals to increase your bookings.

If you do decide to join one of these organizations, make sure to include their membership logo on your business card and in your literature. These logos act as an extra reference for you and speak to your professional status in the industry.

Here is a list of professional disc jockey associations:

United States

National

- The American Disc Jockey Association—www.adja.org
- Global Mobile Entertainers Association—www. globalmobile.org
- Mobile Entertainers Guild of America—www.megadjs.com
- National Association of Mobile Entertainers—www.djkj.com
- National Club Industry Association of America—www. nciaa.com

Regional and State

- Albuquerque, New Mexico Chapter of the ADJA— (currently no website)
- Atlanta Georgia Chapter of the ADJA—www. georgiaADJA.com
- Arizona Chapter of the ADJA—www.azadja.org
- Baltimore Area Disc Jockey Association—www.badja.org

- Bay Area Mobile Music Association—www.bamma.org
- Central Illinois Disc Jockey Association (ADJA chapter)—www.cidja.com
- Chicagoland Chapter of the ADJA—www.chicagoadja.com
- Colorado Professional Disc Jockey Association (ADJA Chapter)—www.cpdja.org
- Connecticut Professional Disc Jockey Association—www.cpda-djs.org
- Dallas/Fort Worth Chapter of the ADJA—www.dfwadja.org
- Disc Jockey Association of Western Pennsylvania—www.djawp.com
- Georgia Chapter of the ADJA—www.georgia.ADJA.com
- Georgia Mobile Disc Jockey Association—www.georgiadjs.com
- Greater Houston Area Mobile Music Association—www.ghamma.com
- Hawaii Chapter of the ADJA—(currently no website)
- Houston Chapter of the ADJA—www.hadja.org
- Indiana Disc Jockey Association—www.ipdja.com
- Iowa Disc Jockey Association—www.prodj.com/idja
- Kansas City Missouri Chapter of the ADJA—www.kcadja.org
- Kentuckiana Area Disc Jockey Association (ADJA chapter)—(currently no website)
- Los Angeles Chapter of the ADJA—(currently no website)
- Maine Disc Jockey Network—www.maineweddingdj.net
- Michigan Disc Jockey Network—(currently no website)
- Mid Atlantic Professional Disc Jockey Association—www.mapdja.com
- Mid-Michigan Disc Jockey Association—www.members.aol.com/mmdja
- Midwest Association of Professional Disc Jockeys—www.mapdj.com
- Minnesota Association of Professional Disc Jockeys—www.mapdj.com
- New Hampshire Chapter of the ADJA—(currently no website)
- New Jersey Disc Jockey Association—www.njdja.com
- New York Chapter of the ADJA—(currently no website)
- New York State Disc Jockey Association—www.nysdja.org
- Northwest Professional Disc Jockey Association—www.spokanedjs.com

- Ocean State Disc Jockey Association—www.osdja.com
- Pacific Coast Disc Jockey Association (ADJA chapter)—(currently no website)
- Professional Association of Disc Jockeys (Los Angeles ADJA chapter)—www.padj.org
- Professional Disc Jockeys of New England—www.pdjne.com
- Sacramento Chapter of the ADJA—(currently no website)
- Salt Lake City Utah Chapter of the ADJA—www.utahadja.com
- San Diego Disc Jockey Association—www.sddja.org
- Seattle and Puget Sound Chapter of the ADJA—www.NWDJ.com
- Silicon Valley California Chapter of the ADJA—www.scadja.org
- Southeastern PA DJ Association—www.sepadjassoc.com
- South Lake Tahoe Chapter of the ADJA—(currently no website)
- Syracuse, New York Chapter of the ADJA—(currently no website)
- Tampa Professional Disc Jockey Association (ADJA chapter)—www.tampaprodjs.com
- Utah Chapter of the American Disc Jockey Association—www.utahadja.org
- Western New York Mobile Entertainers Alliance—www.wnymea.com
- Wichita Kansas Chapter of the ADJA—(currently no website)

Canada

National

- The Canadian Association of Mobile Entertainers and Operators—www.cameodj.com
- The Canadian Disc Jockey Association—www.cdja.org

Provincial

- Alberta Association of Mobile Entertainers—www.aame.ab.ca
- Montreal DJ Association—www.montrealdj.com

England

- South Eastern Discotheque Association—www.seda.org.uk

Online

- Canadian Online Disc Jockey Association—www.codja.com
- Online Disc Jockey Association—www.usodja.com
- United States Online Disc Jockey Association—www. usodja.com

Trade Shows

Attendance at trade shows is essential for DJs who want to keep on top of the rapid changes that continually shape the scope of our business. Disc jockey conventions are action-packed and fun-filled opportunities to grow as a professional.

There are workshops and seminars on a variety of industry-related topics. You can also network with and learn from peers. A visit to the exhibit hall offers DJs the unique opportunity to see, hear, experience, and discuss the offerings of hundreds of vendors in one place. Special parties and social events are a great way to meet and speak with others under highly enjoyable conditions.

On returning home from a DJ convention or other trade show, be sure to organize the business cards you have received and the information you have taken from vendors. Review any notes you have taken, and create an action plan from what you have learned. Make the most of your trade show experience.

Here is a listing of professional disc jockey conventions and cruises as well as other trade shows that may be of interest to disc jockeys:

- America's Band and DJ Conference & Trade Show— www.americasbanddjshow.com
- Audio Engineering Society—www.aes.org
- Billboard Dance Music Summit—www.billboard.com
- Canadian Music Week—www.cmw.net
- DJ Cruise—www.djcruise.com

Fellow DJ and author Ray Martinez gives Stacy Zemon a hug for luck just before her "Dollar$ from Sense" keynote address at the 2004 Mid-America DJ Convention in Louisville, Kentucky.
Photo courtsey www.ProDJ.com.

Vendor booth at DJ Times International DJ Expo in Atlantic City, New Jersey
Photo courtesy *DJ Times* magazine.

- Entertainment Technology Week (Lighting Dimensions International)—www.ldishow.com
- International DJ Expo—www.djtimes.com
- International Nightclub and Bar Convention & Trade Show—www.nightclub.com
- Karaoke Fest—www.karaokefest.com
- Mid-America DJ Convention—www.midamericadj.com
- Mid-Atlantic DJ Convention—www.mapdja.com
- Mobile Beat DJ Show & Conference—www.mobilebeat.com
- Musikmesse/Prolight & Sound—www.musikmesse.de
- National Association of Broadcasters—www.nab.org
- National Associations of Music Merchants—www.namm.com
- Plasa (Professional Lighting and Sound Association)—www.plasashow.com
- WeDJ.Com Convention Cruise—www.wedjcruise.com

In addition to the preceding list, you might also consider attending photographer, videographer, and caterer conventions. The trade associations and conventions information for all these related industries is easily obtainable on the Internet.

Note: Stacy Zemon provides the listings of professional disc jockey associations and trade shows for informational purposes only. She does not necessarily endorse and does not take responsibility for the associations listed or their services. Apologies are made for any accidental omissions or errors.

*What you get by reaching
your destination is not
nearly as important as
what you will become by
reaching your destination.*

ZIG ZIGLAR

10

Records of Success: Inside Secrets from Prominent DJs

Define success on your own terms. Don't let others make the rules for your happiness. Often it's a balance of family life, personal achievement and doing good for others. Don't let other's "shoulds" and "musts" rule your choices. Make up your own mind and set the course for your own life.

SID VANDERPOOL

Owner/DJ Entertainer:
Music Magic Professional Disc Jockey Service, Twin Falls, Idaho
CEO: DJChat.com (The largest online community for DJs in the world)
Co-Founder: Disc Jockey News Network (www.DJNN.com)
Editor: DJZone.com and DJZone.net, online DJ magazines
Author: *The Ultimate DJ Games Book*
Speaker: DJ Times International DJ Expo, Mobile Beat DJ Show and Conference, Mid-America DJ Convention
Writer: *Pro Audio Review* and *DJ Entertainer* magazines
Member: National Association of Mobile Entertainers

RANDY BARTLETT
Owner: Premier Entertainment, Sacramento, California
Company Slogan: "For People Serious About Having Fun"
Producer: The 1% Solution DVD Series and Newsletter
Keynote Speaker: 2003/2004—The Association for Wedding Professionals International
Member: American Disc Jockey Association
Speaker: Mobile Beat Trade Show and Conference, DJ Times International DJ Expo, MAPDJ

Photo courtesy
Randy Santee,
Santees Designer
Images.

How do you succeed and prosper with your DJ service?

Randy Bartlett
The DJ business is just like all others, and unlike all others. It is, in fact, a business, and needs to be run like one. Understanding basic business principles and surrounding yourself with knowledgeable, trustworthy people is a must. I hate bookkeeping, so I have a book-keeper, which frees me up to do the creative side of things, which is what I do best.

Unlike other businesses, purchasing our services is an emotional experience for most clients—especially weddings and mitzvahs. Understanding how to sell based on this knowledge is crucial to your success.

From the start, my company has been almost fanatical about in-house and client evaluations. We track them very closely and use the feedback to improve our services. This has become the single most important ingredient in our success over the years.

How do you run a thriving operation?

Randy Bartlett
It's evolved over the years. In the beginning, it was mostly trial and error, but we paid attention to our mistakes and to what we did well.

Today, we've carved out a niche as the most sought-after DJ company in our market. We made a conscious decision to provide the best-quality service and the best-quality product, and we and our customers both understand that doesn't come cheap. Prospects and clients have heard about or know how we perform, and they seek us out.

We gave up on schools and casual events long ago. While we still perform at an occasional birthday party or reunion, it is almost always for a former client.

I only hire people for our staff who genuinely care. We can train DJs to improve their skills, but we cannot train sincerity—that's an inside job. When one of our entertainers truly cares about a client rather than just about him/herself, or a paycheck, the results show.

Finally, we invest in ourselves. I'm amazed at the number of DJs who will spend $200 on a new light but won't spend any money attending conventions, subscribing to trade publications, attending improvisation classes or comedy workshops, joining a trade association, or reading business books. You had the wisdom to buy this one, so you're on the right track!

Money spent on skills is money wisely invested. Remember, Tiger Woods has a coach, and he practices every day.

How do you market and sell your services?

Randy Bartlett
Marketing and selling are two different things. Marketing is what you do to get your phone to ring. Selling is what you do after the phone rings.

In the beginning, we marketed through *Yellow Pages,* bridal fairs, and placing business cards in wedding shops and party stores—You name it, we did it!

We use a very consultative, no-pressure sales approach. For our company, it's about building relationships and trust. The prospect who calls us just looking for a low price will not likely become one of our clients, and that's okay. Our business model is based on booking the right events, rather than just trying to book as many as possible.

Becoming an excellent salesperson is critically important to the success of your business. If you haven't attended a professional sales school or seminar, make that your number one priority. Why?

Because most businesses that fail, do so because they were unable to effectively communicate their value to potential clients, which is all selling is really all about. It's a skill that can be learned.

How should a new DJ company price their services?

Randy Bartlett
It depends entirely on whether they are new to the art, or just starting a new business. If they are new and "green," the first thing is to work for a DJ company or for yourself because this is the best way to gain experience.

For "newbies," don't oversell yourself and price your services so that you can work every weekend. As you gain experience, you'll need to analyze where your talents fall in relation to other DJs in your market. If you're pretty average, you'll need to price yourself accordingly. If you're above average, you should definitely charge an above average price. You can obtain feedback from peers on this subject by joining a local or regional trade association.

My company charges two to three times the average rate for DJ services in our market. As a result, I believe it is our lead that helps others to earn more when they are ready. You have to decide whether to be a leader or a follower.

Ultimately, this should be a profession that allows you to make a comfortable living, including benefits and a retirement plan. If you don't get there within five years or so, it's time to re-evaluate your business skills or your performance skills, or both.

How much of your annual budget is dedicated to marketing and advertising?

Randy Bartlett
Less than 1%. In the last 10 years, I've spent maybe $500 on all marketing and advertising expenses. Our business is now completely referral. The only money we've spent is our annual membership in The Association for Wedding Professionals International, and the monthly mixer fee.

In what ways are networking and public relations incorporated into your business?

Randy Bartlett

For most companies, this networking and PR is an integral part of their business. My business belongs to the Association for Wedding Professionals International, which has a very strong local membership. I attend about five or six mixers each year, and use my MC skills at their year-end awards show.

We used to do a lot more but as our business has evolved into all referrals over the years, our marketing, networking and PR have dropped to virtually nothing. Because our clients enthusiastically rave about our DJ services to others, we are kept very busy.

How do you integrate customer service into your DJ service?

Randy Bartlett

Customer service has become an overused cliché that has lost most of its real meaning. Everyone thinks they give great customer service. Have you ever heard of a company who says they don't provide good customer service? Customer service is a lot more than returning phone calls or sending thank you notes.

For us, customer service starts with understanding why we are hired. Because we primarily work at weddings, mitzvahs, and corporate events, we realize that it's not just about the music, but about the experience, so we start by learning what sort of experience the client really wants, and then we help them create it.

There's a great test to see how your clients really feel about their experience with you. If they shake your hand and thank you at the end of the night, you've done a good job, and you have a satisfied customer. And you'll spend a fortune on marketing to get another customer to replace them.

But if they hug you, and look into your eyes and tell you that you've given them the best night of their lives, then you've really given them customer service, and they'll do all the work to get another customer for you.

PAUL BEARDMORE
Owner: The DJ Connection, northern Virginia and
Washington, D.C.
Founder: DJ Cruise

Photo courtesy
www.prodj.com

How do you succeed and prosper with your DJ service?

Paul Beardmore
So many of us are good entertainers but failures as business
managers. This, unfortunately, can ultimately lead to the failure of
our DJ businesses. The key to running a successful entertainment
company is to have the ability to transform your excitement and
ability to entertain people into a sound business action plan.

Successful disc jockey services focus on more than just improving their entertainers' performance abilities. They also focus on
marketing their services as well as following the fundamental business principles that allow your service to be profitable.

This is why it is critical for anyone who owns a DJ company to
seek higher educational opportunities at local community colleges
and DJ conventions, as well as networking with other business
owners.

How do you run a thriving operation?

Paul Beardmore
To keep my business thriving, I am constantly looking for ways to
improve my business in all critical areas. The single greatest reason
my company is growing and thriving, is because I have an excellent

network of DJ friends. They are not only great entertainers, but also sharp business people.

Sharing information and ideas with my network of disc jockey peers has been instrumental in the success of my DJ service.

How do you market and sell your services?

Paul Beardmore

I have always felt that using a "soft" approach was better than an aggressive sales pitch. The trick to selling your services is to engage the client with open-ended questions that get them to do as much talking as possible. Once you "break the ice" with a potential client, you can ease into a basic overview of the services you have to offer, without sounding like a used car salesperson.

How should a new DJ company price their services?

Paul Beardmore

A new company needs to be realistic about the value of their services. There are a lot of dynamics that determine the appropriate fee for someone just entering our business.

Let's say the average mobile entertainer in your market is charging $800 for a four-hour wedding. In this case, it wouldn't be appropriate for a new DJ with limited performance skills to charge the average amount in the early stages of her/his business.

Underpricing your services isn't the answer either. If you are just starting out and have a reasonably professional sound system and music library, then charge about 80% of the average price for most DJs in your area. Increase your price each year as you become a better emcee and entertainer.

How much of your annual budget is dedicated to marketing and advertising?

Paul Beardmore

8% to 12% of our overall budget.

In what ways are networking and public relations incorporated into your business?

Paul Beardmore
I am a member of the local Chamber of Commerce as well as a local wedding professionals association. Networking with other wedding professionals is an important part of public relations. This is because it is important for *people* in my community to look upon our company in a positive manner.

Even after about a quarter of a century in the business, I never take for granted the hard-earned reputation we have in our area. Stopping by and saying "hello" to a local photographer, venue manager, etc., goes a long way to maintaining a favorable reputation of my DJ service.

How do you integrate customer service into your DJ service?

Paul Beardmore
Customer service is an area that many DJs fail to deliver to their clients. My company focuses on customer service both before and after events. Proper planning and organizing is critical to the emcee and DJ's ability to provide a polished performance. It is important to go the "extra mile" and do what it takes to make a client happy.

ROBERT DEYOE
President: Desert DJs, Tucson, Arizona
Founder/Director: DJ Camp
Speaker: DJ Times International DJ Expo
Recipient: 1999—DJ Times International DJ Expo
"DJ of the Year for the Best New Game" category
award; 1997—DJ Times International DJ Expo;
"Top 10 DJ Entertainers" award; 1996—DJ Times
International DJ Expo "Top 10 Wedding DJs" award

How do you succeed and prosper with your DJ service?

Robert Deyoe
It's best to be very ethical and to treat all of your clients fairly. If somebody is not happy with a job you did for them and you don't respond, it will come back to haunt you.

The only companies that have ever done well in our area are those that market themselves, and continually shake hands with every new catering director and party planner.

You have to be able to back up what you say with really good DJs. If you want to be successful, you need a really good product. Your disc jockeys have to be well trained. Let everyone in town know you're absolutely the best and you'll guarantee it!

If you want to get into this business, go into it for the right reasons. Do it because you feel passionate about being an entertainer, and are committed to succeeding as a business owner. Our mission statement talks about providing clients with the best possible service, and that's why we're here. We would certainly like to make a profit, but we also want to provide the best possible service.

How do you run a thriving operation?

Robert Deyoe
I think part of it was the timing. We made a huge commitment about twelve years ago to be #1. There was a company about three times our size that was very well known. Our goal was to emulate them and do what they did even better. It took a couple of years, and they "blinked." We quickly caught up to them and they ended up going out of business.

We have excellent equipment and talented people whom we treat really well. I have strong marketing ability and my wife, Jennifer (one of my former DJs), is skilled at creating marketing materials from my ideas.

How do you market and sell your services?

Robert Deyoe
In addition to placing a small ad in the *Yellow Pages,* we primarily market our services using a three-color brochure. Our direct-mail package includes copies of articles and interviews that have been conducted with our company. We like everything we send out to look really, really sharp. Additionally, prospects are invited to stop by the office and visit us. We have six videos for them to view, including two that are specifically geared for bridal couples.

How should a new DJ company price their services?

Robert Deyoe
Before starting our DJ business, we conducted research on the competition in our area, then priced ourselves 10% to 20% below them. Initially, that was the draw for people who had never used our service to try us out. We were confident that they would be ecstatic about our services.

Over the years we have slowly raised our rates to be commensurate with our competition, eventually making our rates a little bit higher. I recommend the same process for any new disc jockey company. In addition, you can ask local catering managers and party planners for their feedback on pricing.

How much of your annual budget is dedicated to marketing and advertising?

Robert Deyoe
About 6% of gross. If I were brand-new it would have to be a lot more—20% to 30%.

In what ways are networking and public relations incorporated into your business?

Robert Deyoe
Professional networking and PR is woven into everything we do. This ranges from speaking to event guests who are potential new clients, to hiring new entertainers or office staff.

How do you integrate customer service into your DJ service?

Robert Deyoe
From the moment a new member of the Desert DJs team is hired, they are reminded that *everyone* deserves great service and to be treated respectfully—even the obnoxious "play more Skynard" guy at a company party. I do my best to role-model this behavior for my staff. Keeping the "Golden Rule" in mind when interacting with people, naturally results in more business.

BRIAN DOYLE
Co-Owner: Denon & Doyle Disc Jockey Company,
Concord, California
Speaker: DJ Times' International DJ Expo

How do you succeed and prosper with your DJ service?

Brian Doyle
We keep asking ourselves and our clients why they would choose us over the competition. We think we do a great job, but in the end, it is really what our clients think that matters.

After performing at over twenty thousand events, my company still constantly learns from each occasion. We try to keep our ego out of the equation and always learn from our missteps.

How do you run a thriving operation?

Brian Doyle
Remember you are in business first and a DJ second. Watch cash flow (you really do not need every new DJ toy out there). Also, treat your staff like partners in your endeavor.

My company has never used non-compete agreements. We believe that if someone really wants to leave us, then we really do not deserve to keep them. Thankfully, in twenty-one years we have never lost anyone that I have regrets about.

By demonstrating trust in your staff, they will repay you tenfold in loyalty and pride in your company. Be aware that your employees will often make decisions you feel you could have made better. Give them leeway. In the long run it will be worth it, and you will find more time for your other life pursuits.

How do you market and sell your services?

Brian Doyle
When our DJ business first started, large amounts of money were spent on advertising. This helped build name recognition for us. Once this was accomplished, we concentrated on building our referral base. Getting on a banquet locations preferred vendor list is a gold mine so we have put quite a bit of effort toward achieving this goal. About 90% of our business now comes through direct referrals.

How should a new DJ company price their services?

Brian Doyle
Until you have made a name for yourself, and have all the right marketing and sales arts perfected, you can really only succeed on price. Find out what your competition is charging and price yourself a couple hundred dollars under that amount.

About 20% of prospects will hire the cheapest DJ no matter what. As you improve, slowly start going after those potential clients who will pay a little more for the assurance of better quality.

A friend gave me a great piece of advice. He said that we should be booking about 40% of our inquiries. If we are booking lower, our prices were too high. If we were booking higher, then that our prices were too low.

How much of your annual budget is dedicated to marketing and advertising?

Brian Doyle
About 10% goes toward marketing and advertising.

In what ways are networking and public relations incorporated into your business?

Brian Doyle
We have a large showroom at our facility. Photographer and catering network associations are invited to use it at no charge.

This public relations tactic has increased our vendor referrals enormously.

How do you integrate customer service into your DJ service?

Brian Doyle

Customer service is what it is all about with us. We are always asking our customers and other professionals how we are doing—and we listen to them. If for some reason we are not comfortable with an event, we don't take it on. Frankly, we would much rather turn down business then take a chance that our services receive a bad review on some Internet chat board.

JEFF GREENE
President: Party Time DJs, Davie, Florida
Winner: National awards and recognition in the international DJ community, including "Best Party DJ in America"

Photo courtesy Debbi Greene, Party Time DJs, Inc.

How do you succeed and prosper with your DJ service?

Jeff Greene
Planning and hard work. The days of simply answering the phone and taking a booking are gone. Clients, driven by the Internet, want to know very little. They want to research on their own and just get a price. But we are not selling hard goods that can be had at any retailer. We are selling personal life experiences. Memories. So it is our responsibility to rise above the price shopper and encourage our prospective clients to meet face to face in order to determine if we want to do business with them. That's right, it should be you who decides if this client is a good fit. Then plan a strategy to educate your clients about your services and learn their needs. You'll always find the right match.

How do you run a thriving operation?

Jeff Greene
I believe very heavily in running my businesses with a "hands-on" approach, and am very aware of what the competition is doing. I also believe in constant training to improve performance. That's why I conduct in-studio trainings for my staff.

How do you market and sell your services?

Jeff Greene
Word of mouth would have to be number one. I only use the free business listing in the *Yellow Pages*. We used to be the biggest advertiser, but now we do a hell of a lot more business than we did then. That's got to tell you something. Today we advertise in direct marketing sources. For example, wedding and bar mitzvah shows, special events, and advertising guides. Performing on stage at showcases allows many prospects to preview our services at one time.

How should a new DJ company price their services?

Jeff Greene
You need to decide whether to establish your business as a full-time or part-time operation. You have got to put together a budget and know your fixed and variable costs. Then, charge the highest dollar amount that you believe you can get.

For example, if the average DJ service in your area charges $400 for a four-hour party, and you know that your entertainment abilities are better than average, then why not charge $600 for the same number of hours?

Initially you may not book as many gigs; however, the clients who do book you are those who really appreciate the difference. They are also likely to refer you more often.

There is a certain segment of the population that believes that the highest price means the best service. You know what? They're right!

How much of your annual budget is dedicated to marketing and advertising?

Jeff Greene
Too much; roughly 15% to 20%.

In what ways are networking and public relations incorporated into your business?

Jeff Greene

The public does not want to hear what Party Time DJs did for whatever charity or organization. Sure it fills newsprint, but all the publicity we've gotten has not paid any bills. Instead, use any publicity as part of a press kit to give to those you wish to do business with. When you network, use your press kit as a follow-up to help get your foot in the door with that caterer.

How do you integrate customer service into your DJ service?

Jeff Greene

Customer service is at every level of what we do. We are in a personal service business and if you want to succeed in this business, learn to serve your customers' needs.

Photo courtesy
St. John Photography.

KEVIN HOWARD ST. JOHN (A.K.A. KEVIN HOWARD)
President: The Howard Group, Inc., d.b.a. Sounds Unlimited, Premier Entertainment, The Fantasy Casino, Events Northwest, E4 Events, Grad Nights, St. John Photography, Seattle, Washington
Company Slogan: "We celebrate life!"
Founder: DJ of the Year Awards (the first national award program for professional disc jockeys), Seattle's Table Prepared Meal Program
Executive Board Member: National Association of Catering Executives, Pacific Northwest chapter; NACE National Convention Co-Chair; Food Lifeline; Seattle Shakespeare Company; Character Core—Kids First
Board Member (past and/or current): American Disc Jockey Association; NAPD; Meeting Professionals International; International Special Events Society; World Entrepreneurs Organization; Young Entrepreneurs Organization
Recipient: 2003 and 2005—Voted Seattle's Best DJ Company by *Seattle Magazine*; Business of the Year Award for Washington State from the Better Business Bureau and U.S. Bank

How do you succeed and prosper with your DJ service?

Kevin Howard
I constantly search for great people to bring on to my team. Since I no longer DJ personally, it is far more important for me to find the right people who can continue to provide the leadership, and insight into our industry. Frankly, I no longer want to know what the latest "got to have" song is for the high school dances. What I do want to know now is, what I can do to help inspire and motivate my team. They are the ones that win the industry awards and the praise from our clients. They are the ones that make me look good.

How do you run a thriving operation?

Kevin Howard
Ten years ago I would have said (and I did), "Work really hard!" Today I know that the answer is, "Work really smart!" My business

exists whether I'm there or not—kind of like a house. A house is still a house, whether or not you're inside of it. You can go away for months and come back and it's still there standing. That's the way a business can be too, if you've built it correctly. My companies operate with a clear set of values—integrity, honesty, and doing the right thing.

How do you market and sell your services?

Kevin Howard

As intelligently and carefully as possible. The days of throwing money at every advertising opportunity are gone. We are constantly measuring our return on investment for all of our promotional and advertising campaigns. While we do aggressively advertise in several mediums, we turn down far more than we engage. For the last few years, bridal shows have dominated our efforts. My companies engage the marketplace with a strong out-bound sales force, and our website increasingly has produced a good ROI. Concurrently, we've seen a reduction in the effectiveness of our *Yellow Pages* campaign and direct mail programs.

How should a new DJ company price their services?

Kevin Howard

People choose products and services based upon three factors: Price, Quality and Service. In order to be successful you should try to excel in any two of those three categories. You can't be the best in all three and remain in business—it's simply not profitable. It's up to you to choose which two categories are the most important for you, go for the best service and quality but with a high price, or be the low price leader and sacrifice either quality or service. In many cases a new DJ simply won't have the experience to be able to offer the best quality and therefore, it is in the area of price and service where they can be the most effective competitor.

How much of your annual budget is dedicated to marketing and advertising?

Kevin Howard
10% to 20%.

In what ways are networking and public relations incorporated into your business?

Kevin Howard
Networking is a huge part of our marketing effort. More than half of our bookings come from industry referrals for our companies. There are a lot of good DJs out there, so we "go the extra mile" to ensure that the people in the events industry get to know us well on a personal level. We believe that people do business (including referrals) with people that they know, like, and trust. You have to work at that if you're going to be successful in this business.

How do you integrate customer service into your DJ service?

Kevin Howard
Customer service is our business. Everyone on our staff understands that it is an honor when a client entrusts us with their once-in-a-life-time event. When you look at it that way, it's much easier to accept the challenges that are inherent in our industry. Difficult clients become more easily understood, unanticipated issues become more manageable, and everyone on the team is willing to put forth just a little more effort to make sure that every client has a truly remarkable event.

In my company, this means having someone available to answer client questions ten hours a day, seven days a week. It also means that we have a back-up DJ every weekend in case there's an emergency. On a more subtle level, providing great customer service means anticipating what a client may need even before they do—let alone verbalize the need.

Photo courtesy
Tony Montrelli,
Tony Montrelli
Photography.

RAY MARTINEZ (A.K.A. RAY MAR)
Disc Jockey/Emcee: Ray Mar Productions, Goodyear, Arizona
Company Slogan: "Performance Beyond Expectation"
Co-Founder: Crossmix
Author: *Performance Beyond Expectation*
Member: American Disc Jockey Association National Board, ADJA Arizona chapter, ADJA Southern California former vice president
Recipient: 2005, *Mobile Beat* Magazine Editorial Advisory Team; 2003, American Disc Jockey Awards "Male Entertainer of the Year," ADJA Hall of Fame inductee; 2002, ADJA Lifetime Member Award; 2001, 2002, ADJA Southern California chapter Entertainer of the Year; 2000, "Michael Butler Humanitarian of the Year Award"; 1999, Recognition from former President Ronald Reagan, the U.S. Congress and California Governor Pete Wilson County for 25 years of entertainment, Proclamations for the same achievement from the California cities of Los Angeles, Anaheim, and Yorba Linda; 1994, ADJA Excellence Award for contributions to the Southern California chapter; 1999, Appreciation Award from American Disc Jockey Awards Show.

How do you succeed and prosper with your DJ service?

Ray Mar

Being an entertainer for over thirty years, believe it or not, it wasn't until I was well past my twentieth year that I finally realized that I could not run my business without putting God first. I firmly believe this because it is He that gave me the ability to perform. I have faith that He will also provide for *all* my needs.

Once I put my priorities in line, God first, family second, and business third, I have prospered ever since. I just wish I had done it much sooner.

How do you run a thriving operation?

Ray Mar

I believe in order to run a successful business you must demonstrate four essential traits: Honesty, consistency, integrity and character.

Display honesty by not trying to scam anyone. Demonstrate consistency by offering the same fees to all new clients. Display integrity by following through on your promises. Finally, display character by saying what you mean and meaning what you say. In other words, don't make promises you can't keep.

How do you market and sell your services?

Ray Mar

I market my company by networking with photographers, videographers, florists, catering directors, and banquet managers. Building personal relationships with these people is the best marketing tool you can have. I also send anniversary cards to my brides and grooms, and then send them an updated business card yearly. In addition, I use my website to market my services.

I like to read through magazines to see how other companies do their marketing. I scan the ads for buzz words that apply to our industry for possible use in my marketing materials.

How should a new DJ company price their services?

Ray Mar

Skill, talent, and experience are major factors that should go into pricing your DJ service. For those just starting out, $125 to $150 per hour is a good rate. I agree with my friend and colleague, Mark Ferrell, that experienced professionals should warrant an average $1200 fee for a four-hour booking.

I believe this is a fair rate because on the average, most part-time DJs across the country charge $75 to $100 per hour. They do so with the mindset that this is more than the hourly rate they earn at their regular jobs. By maintaining this mindset, they will never achieve a higher income. A better approach would be to start your rates out a little bit higher than average for your area.

This helps raise the standard fees for our industry; however, don't just become a DJ for the money because you won't enjoy the work, which will be apparent to clients, and will turn you into a commodity. Passion and dedication are the keys to making more money.

How much of your annual budget is dedicated to marketing and advertising?

Ray Mar

My budget in California was about 10% because I was well established in the marketplace. However, now that I have moved to Arizona and need to establish my business presence here, I plan to increase my budget to 15% to 20%.

In what ways are networking and public relations incorporated into your business?

Ray Mar

I keep my referral contacts happy by occasionally taking them out to lunch or dinner. I remember them with a card during the holidays and call them once or twice a month to keep in touch and express my appreciation to them.

Again, I can't stress strongly enough, the importance of building relationships. Get to know the people who are helping you by taking a genuine interest in them. This will go a long way toward helping your business grow and prosper.

How do you integrate customer service into your DJ service?

Ray Mar

One facet of customer service that I regularly incorporate, is taking a genuine interest in my clients by listening to them, and finding out what their needs and desires are. Listening is vital because they will tell you what they want. I don't try to hard-sell anyone.

I always let clients know that my services include not only entertaining and playing music, but also providing referrals to other vendors.

If you serve your clients from your heart without expectations, then you will reap the financial rewards. You will also feel the joy that comes with knowing you helped someone's event be successful.

This is how client loyalty starts and your referral network will flourish.

PETER MERRY
Wedding Entertainment Director: Last Dance
Entertainment, Ladera Ranch, California
President: 2004/2005, American Disc Jockey
Association
Recipient: 2005, "Peter Merry Leadership Award"
from the American Disc Jockey Association
Speaker: DJ Times International DJ Expo, Mobile Beat
DJ Show and Conference

Photo courtesy
Peter Merry,
Last Dance
Entertainment.

How do you succeed and prosper with your DJ service?

Peter Merry

If your definition of success is to create a business that runs itself and can be sold at some point in the future, then developing a strong multifaceted organization should be your primary goal. If working at as many events as possible because you just love being a DJ is your definition of success, then increased profits might not be your driving motivator.

If building demand for your unique talents and garnering the best price possible for your services is your definition of success, then creating a quality performance and solid sales skills are the most important factors.

Generally speaking, giving your customers more than they expect and never treating their events as "just another gig" will always render increased referrals while also building a solid reputation for dependability and quality.

How do you run a thriving operation?

Peter Merry

Put your client's wants and needs ahead of your own, and they will pay you handsomely for it. Do what other DJs in your market are unable or unwilling to do for their clients, and new clients will beat a path to your door. Treat this vocation as a full-time business and career, and soon it will be (treat it like a hobby and it always will be). Find new and innovative ways to do the same old things and don't get stuck in a rut.

Being successful in any entertainment field is like running up the down escalator. Once you stop running, you are already losing ground.

How do you market and sell your services?

Peter Merry
Currently, I don't do any paid marketing. I also haven't paid for any advertising in over four years. My performance is my best marketing tool.

Focus on developing a uniquely personalized performance for each of your clients, and you will get more word-of-mouth referrals than you can handle.

I sell my services by doing just that, selling my services, not my gear or my music collection. I sell my services by taking the time to find out exactly who my clients are, and exactly what they do and don't want in their wedding entertainment. Then, I show them how I can make that happen for them in their homes during an initial, one-hour consultation.

How should a new DJ company price their services?

Peter Merry
When I started my DJ business in 1995, I checked out what other DJs in my market were charging by calling their *Yellow Pages* ads. I have since learned that this approach is very dumb for a variety of reasons.

First, most of the high-end DJs in my market don't advertise in the *Yellow Pages* because the ads only attract price shoppers. Thus, the figures I was getting quoted for DJ services were skewed toward the lower end of the scale.

Second, I falsely assumed that all of the companies listed in the *Yellow Pages* were successful and making a good profit, so I never bothered with creating my own business plan. I figured if they are charging that much, then that must be a reasonable price.

I have since learned that a reasonable price is much better determined by investigating how many weddings (or plug in your own favorite events) occur annually in my region, and how much people are spending on those weddings (and on related wedding services).

I also discovered the power of surveying my clients to find out if they thought my services were worth more than I was currently charging. They consistently tell me that they have received far more value then cost.

I currently do about forty-five wedding receptions a year, plus an additional ten to fifteen other types of events. The current fee for my wedding services is approximately $2,850 for a reception only and $3,600 with ceremony coverage. I work solo with two to four speakers, and no lights or props.

Don't ever base your price entirely on what other DJs are charging. If they're going broke at their rates, you will too. 15 percent to 20 percent of your client's total budget is a fair trade for carrying 80 percent of the responsibility for the success of their event. To learn more about this, I would suggest you purchase the CD seminar series "Getting What You're Worth" by Mark Ferrell at www.djsecrets.com.

How much of your annual budget is dedicated to marketing and advertising?

Peter Merry
Less than 1%. Website hosting and printing new business cards have been my only ongoing marketing and advertising expenses. I have invested my top priority into developing personalized entertainment for my clients, which has resulted in attracting only word-of-mouth clients for the last 4 years. The best part is that word-of-mouth clients are much easier to sell to than cold calls generated by an advertising campaign.

In what ways are networking and public relations incorporated into your business?

Peter Merry

My primary resource for networking has been my involvement with the ADJA. Our local chapter meetings have put me in touch with like-minded DJs with whom I have been able to refer overflow clients back and forth. Networking with other ADJA members nationwide has expanded my collection of creative performance ideas, as well as gaining new perspectives on differing business models for success in the DJ industry.

How do you integrate customer service into your DJ service?

Peter Merry

As often as possible. I return calls and e-mails as quickly as possible—and always on the same day. I offer many options from which a client can choose so they can customize their ideal reception rather than feel that they are getting a cookie-cutter performance. I meet with clients at least twice in advance to help plan their reception. If I am involved in providing PA support for their ceremony, I attend the rehearsal, and more often than not, I am also the one who directs the rehearsal for them. I contact their other vendors in advance and provide them with a copy of the reception agenda. I show up extra early so I can assist anyone else who may need my help once I have completed my setup.

MATT PETERSON
Owner/DJ: Peterson Productions Disc Jockeys,
Northampton, Massachusetts
Company Slogan: "Your Source with the Music Force"
Director: Professional Disc Jockeys of New England
Member: National Association of Mobile Entertainers

Photo courtesy
Bob Blanchard,
RJB Photography.

How do you succeed and prosper with your DJ service?

Matt Peterson
A lot of hard work, determination and let's not forget about the endless phone calls, e-mails and very late nights.

When you first start out, it is very important to build a reputable business that you can be proud of. In the long run, you must remember to treat all of internal and external customers, prospects and associates with sincere respect. Be honest not only with them but with also with yourself. Do not promise the moon and stars, and then not deliver them.

How do you run a thriving operation?

Matt Peterson
I treat all of my clients with the utmost respect and I strive to obtain their total satisfaction. I accomplish this by providing superior customer service, excellent musical programming at events, and by paying total attention to a client's specifications. In addition, I always try to stay informed of new and innovative ways to improve on my services and professionalism. I take a great deal of pride in what I do, and believe it shows in my performances.

How do you market and sell your services?

Matt Peterson
You must first decide what type of entertainer you would like to become, and then determine which market you would like to appeal to. When you are clear on your niche, then cater all of your marketing and sales efforts to this target market. My best results have been through local bridal guides, personal referrals, and most recently, my company website. I am always on the lookout for new ways to advertise. Don't underestimate the power of networking and the Internet.

How should a new DJ company price their services?

Matt Peterson
If you are just starting out, be honest and ethical when setting your prices. Don't think that you can automatically charge the same amount as an experienced professional. You must first pay your dues and get some experience under your belt. Do some research to determine what fees your demographics will bear.

How much of your annual budget is dedicated to marketing and advertising?

Matt Peterson
When starting out, you must be willing to risk a pretty big chunk of your profits for advertising. Eventually, your most effective and least expensive advertising will come from word-of-mouth referrals, and that little marketing billboard called a business card. As your business grows, you will find out what works best for you and your company. When it comes to marketing and advertising, I have found that I usually get what I pay for.

In what ways are networking and public relations incorporated into your business?

Matt Peterson
This is the main focus of our business. I take a great deal of pride in the quality of the services I provide. I also always remember that without customers, I wouldn't have a DJ company. I express my appreciation to clients by thanking them for choosing my services. Try to develop relationships that will last a lifetime.

How do you integrate customer service into your DJ service?

Matt Peterson
When talking to a client, I always remind myself that their event is a special occasion, and that they are personally inviting me to join in on the festivities. While it is true that I have been hired to perform for them, I would like to think that I have been selected as a friend or a member of their family. I am a person who enjoys being in the company of others, and paid to party. Who could ask for anything more?

As DJs, we must always put our customers' needs first and treat everyone with respect, kindness and appreciation. After all, you never know who your next client will be.

J. R. SILVA
DJ, VJ, KJ, Performer, Producer: SILVA Entertainment, Orlando, Florida
Company Slogan: "Excellent Entertainers Everytime!"
Writer: *DJ Entertainer* magazine
Speaker: DJ Times International DJ Expo

Photo courtesy
Derek Smith,
Sunshine
Photographics.

How do you succeed and prosper with your DJ service?

J. R. Silva
You can succeed and prosper by being an extraordinary entertainer. I believe that Talent is Everything, and that DJ entertainers should be self-motivated and able to manage their companies. Good business practices lead to a filled calendar, continuous revenue, and rewarding work.

On the performance side, succeeding means that audiences are having fun being entertained by what you are doing. If you see an opening to do something dynamic, do it with great enthusiasm, especially if you know it's going to please the audience.

Paying attention to what my clients are asking for or anticipating their needs has helped me grow my entertainment company. If you understand your market and know your client's expectations you can't go wrong.

It's clear that our industry and the entertainment business are constantly moving forward, so one must be able to adapt and embrace change. I've found that keeping up with DJ friends in other parts of the country has helped me take advantage of opportunities where both my business and performance skills can grow.

Lastly, it's very important to be a team player when you are at the event so that you will make good contacts and be welcomed back time and time again.

How do you run a thriving operation?

J. R. Silva
I've always tried to be the leader in the market by being the first to bring in new services. I attend DJ trade shows, pay attention to my customers' needs, and create detailed systems for my staff to follow. We excel in theme parties and social events. Weddings are and will always be an important part of my business, but my staff and I also look forward to doing holiday parties, homecoming and prom dances, bar and bat mitzvahs, and company picnics. We love the work and the versatility that it demands.

Staffing is critical because the people who work for you are carrying your name and reputation on their shoulders. I only hire or subcontract performers who have their act together, are team players, can follow my direction, and care as much about the client's expectations as I do. New entertainers spend four to six weeks shadowing our DJs at events so I am sure they are up to speed. Then, they can start adding their own creativity to the mix.

How do you market and sell your services?

J. R. Silva
A lot of it is good old-fashioned networking. For weddings, we have had good results with websites like www.theknot.com. I rely on personal consultations, great brochures, and targeted postcards, newsletters and e-mails to different markets. Video is a fantastic tool for showing a customer exactly what (and who) they're buying. That's why I have specialized videos for weddings, proms and bar and bat mitzvahs, which I either show in-person or mail to potential customers.

How should a new DJ company price their services?

J. R. Silva
Research your market and know what your costs are so that you can be competitive. The money you are earning must be worthwhile, pay the bills, and leave something left over to be worth your while. If your business experiences cash flow problems,

that's usually a sign that something is off kilter and needs to be remedied.

To increase their bottom line, many DJs offer add-ons like lights, novelties and video. Greater showmanship can also allow a top performer to command a higher price. Naturally, Saturdays and holidays always command more money. In Orlando, Florida, there are three pages of mobile DJs in the phone book and countless others who are not listed. Smart shoppers who understand that experience and talent are more important than saving fifty bucks, seek the performer who is the "best-fit" for their event. Price will always be a concern, but value is measured in the eyes of the client.

My advice is to find a niche in your marketplace and price your services accordingly.

How much of your annual budget is dedicated to marketing and advertising?

J. R. Silva

I spend about $5,000 a year even though I'd like to spend more and have marketing campaigns that I plan to further develop. My advice is to find the budgetary balance between reaching new and existing customers.

Yellow Pages advertising has lost its luster. Brides and savvy event planners jump onto the Internet to find their entertainment. We have a bold listing under three headings in the *Yellow Pages* and that's it. By the time they reach "S" for Silva Entertainment, they are price shopping. Our website and individual performance videos are our best promotional tools. My future plans include making www.silvaentertainment.com my virtual office online. This will enable clients to see all that we have to offer from the comfort of their home or office.

In what ways are networking and public relations incorporated into your business?

J. R. Silva

I prefer to work closely with our clients, but it isn't always possible so we have systems in place that enable us to gather information online or via fax. My staff and I help meet our clients' needs by asking a lot of questions. That's how we learn about their goals, wishes and expectations. We use the information gathered to customize our events in the manner the client desires. This makes everyone happy including our client, their guests and even the banquet manager. Successful events always create great public relations for Silva Entertainment.

How do you integrate customer service into your DJ service?

J. R. Silva

We are all in the customer service business. Some of us are just able to see things through the customers' eyes more clearly than others. It's all about accountability really. For example, if you say you are going to send a brochure or a proposal, find a certain kind of song or lead a certain kind of custom, you need to deliver the level of service you are promising.

I teach my staff to be attentive and accommodating to a client's wishes and guest requests. We send both "thank you" notes and report cards out a week after the show to follow up and get client feedback. I also send production assistants out to events, to make sure things are running smoothly. In addition, we videotape our performers so they can see themselves in action. This really pays off in terms of being able to stress the elements of good showmanship and improve everyone's performances.

RANDI RAE TREIBITZ (A.K.A. RANDI RAE, "THE MITZVAH MAVEN")
President/Emcee/DJ: Major Productions, Inc., The Social Butterfly, Randi Rae Entertainment, Edison, New Jersey
Speaker: DJ Times International DJ Expo

How do you succeed and prosper with your DJ service?

Randi Rae
Run a legitimate business and don't be greedy. Continue learning about our trade and act like a professional. Start small and do not over-buy equipment. Stay focused. Treat your DJs well. Remember the client is always right. Bear in mind that when you are out doing a job, you're not just representing yourself, you're representing an entire industry.

How do you run a thriving operation?

Randi Rae
I believe my success is due to a combination of my performance personality, sincerity, and a total commitment to my clients. I work twenty-four hours a day and love performing in front of people. I wouldn't change my career for anything (except maybe a TV talk show).

How do you market and sell your services?

Randi Rae
My entire business is referrals. I do bridal shows where I am the only DJ, and I am able to perform live. I have a "two-foot rule." Anyone I come within two feet of is a potential client, and therefore

should know what I do for a living. I also do a lot of public relations work during a party by shaking hands and meeting people.

How should a new DJ company price their services?

Randi Rae

Price your services competitively, based on your experience. Do not overprice or underprice your entertainment fees. These practices do the DJ industry an injustice, not a service.

STACY ZEMON
President: Stacy Zemon Entertainment
Company Slogan: "Bringing the *Party* to Your Event!"
Co-Founder/Partner: The Radio Mobile DJ Group,
DJ Camp®
Consultant: DJ companies in the US and Europe
Author/Writer: The Mobile DJ Handbook 1st &
2nd edition, *Manual del DJ Movil* (Spanish Translation),
The DJ Sales & Marketing Handbook:
DJ Times magazine
Reporter: Disc Jockey News Network (www.DJNN.com)
Producer: Event fundraisers for the Muscular
Dystrophy Association, the Cystic Fibrosis Foundation,
the Variety Club, various nonprofit organizations that serve the homeless and
AIDS patients.
Speaker: DJ Times' International DJ Expo, Mobile Beat Show & Conference,
Mid-America DJ Convention
Judge: International DJ Expo Disc Jockey of the Year competitions.
Artist Representative: American DJ lighting, PVDJ audio, Promo Only CDs
and DVDs
Career Affiliations: American Disc Jockey Association, ADJA Leadership
Council; American Marketing Association, Advertising Club, Association of
Independent Video and Filmmakers, Association of Production Professionals.
Education & Professional Development: Stephen Covey's "First Things First"
for Managers; Pagett-Thomson's "Knock Your Socks Off" Customer Service;
Landmark Education's Forum, Advanced Course, Self-expression & Leadership,
Beyond Fitness; Lifespring; Insight; Silva Mind Control; Neuro-linguistic
Programming.
Media Quotes: "A veteran of the industry who has contributed much to its
evolution" - DJ Times magazine, "One of the most successful women in the
DJ profession" - Mobile Beat magazine, "A knowledgeable source on being a
professional entertainer" - DJ Times magazine.

How do you succeed and prosper with your DJ service?

Stacy Zemon

It's different now than it used to be. In the 1990's I founded a full-
service, multi-system operation that had twenty-three DJs working
for us. We were partnered with a radio broadcasting company and
had commercials regularly aired to bring in prospects. This strategy quickly made us the #1 in the marketplace.

Today I am a solo operator with Stacy Zemon Entertainment.
Because of my various entrepreneurial enterprises I only want to do

three to four gigs a month. To accomplish this I do subcontracting work for a large multi-system operation in my area as well as doing my own bookings. I have a strong USP, which allows me to garner a very good fee for my services.

How do you run a thriving operation?

Stacy Zemon
I use the S.M.A.R.T. approach to goal-setting and am constantly on the lookout for ways to improve. I have an excellent network of friends and business associates from whom I constantly learn. I also read a lot of success, sales, and marketing-related books, as well as watch DVDs produced by the top people in our industry.

When it comes to my clients, I "go the extra mile" to provide exceptional customer service. I also treat each event as if it is of extreme importance and work hard to provide top-notch entertainment. People regularly come up to me and say, "you look as if you're having fun!" I reply, "I am... and I'm glad it shows!"

How do you market and sell your services?

Stacy Zemon
I regularly talk to people I meet about what I do for a living and hand out a lot of business cards. I am constantly networking and asking for referrals. My company is also listed on several DJ websites. In addition, my booking agent pre-sells me to prospects over the phone. When I speak with potential customers I use a soft-sell approach that is based on fulfilling their needs rather than touting my accomplishments and abilities.

How should a new DJ company price their services?

Stacy Zemon
I highly recommend being an apprentice at a disc jockey service or attending a DJ school before calling yourself a "professional disc jockey" and entering the marketplace. You can obtain feedback from peers on your readiness by joining a local or regional

DJ association. Don't even think about doing weddings or Mitzvahs until you have plenty of experience "under your belt."

As a "newbie," your pricing structure should be roughly 75% of the going rate in your marketplace. Increase your price each year as you become a better entertainer and your business is established.

How much of your annual budget is dedicated to marketing and advertising?

Stacy Zemon
Very little, but I am in a different position than most of my peers because I am a published author, artist representative, and member of the American Disc Jockey Association. I rely heavily on PR and "spin" this into closing sales in most cases. Also, because I subcontract my services, I don't have to incur the expenses related to making my telephone ring with prospects.

In what ways are networking and public relations incorporated into your business?

Stacy Zemon
How aren't they? I talk to peers from all over the US and Europe on a regular basis—over the phone, through e-mails, and through IM messaging. I also attend DJ conventions, read trade publications, network with other DJs, and have mentors and a coach because I am deeply committed to on-going personal and professional improvement. If you want to be REALLY successful, try this approach and see where it gets you!

I take the time to build rapport with banquet managers and caterers at functions. They are generally impressed when I hand them an itinerary of the order of events. At the end of the gig I ask them if they liked my performance. When they say, "yes," I thank them, hand over a few business cards, and ask for referrals.

Press releases from the companies I represent have given me great PR. This helps me both with booking clients as well as giving me credibility in my DJ-related business endeavors.

How do you integrate customer service into your DJ service?

Stacy Zemon

I genuinely care about it, and say "please" and "thank you" a lot. I want to give excellent customer service and performances, so I authentically treat every prospect and client as if they are very important to me. I also convey my desire to do everything possible to make someone's event everything she or he has dreamed of through preparation and performance. Then, I follow through on all of my promises. When I get a hearty handshake or a hug at the end of an event, I know I've succeeded with my goals. I still send out client evaluations after every performance and use the feedback to improve my services.

I have learned the skills of being a very good listener and handling situations when people are upset with calmness and professionalism. Is this trying at times? Oh yeah! However, the end result is almost always worth the effort.

I truly believe that serving clients from my heart, not my ego, has been one of the greatest contributors to my success, both financially and in terms of the joy I have experienced.

Sample Materials

Business cards design. Courtesy, Dowdle Design—www.dowdledesign.com.

Front

D.J. Peace (Mark Thomas)

800-430-4487
626-793-1877

www.awesome-entertainment.net
peace@weddingwarehouse.com

ASK FOR:

Back

Pure Energy Productions

ENTERTAINMENT FOR THE MILLENNIUM

Disc Jockeys · High Tech Lighting · Digital Sound Systems
Karaoke · Lasers · Video Walls · Interactive Games · Rentals

Office: (707) 778-9249   Voice Mail: (707) 491-6681
www.pureenergyproductions.com
P.O. Box 751558 Petaluma, CA 94975

Front

STACY ZEMON
ENTERTAINMENT

"Bringing the *Party*
to Your Event!"

Stacy Zemon
DJ/Emcee

610.747.0704
djstacyz@aol.com
www.stacyzemon.com

Front

MEMBER

THIS WOMAN CAN HELP MAKE YOUR
SPECIAL OCCASION A **HUGE** SUCCESS!

AUTHOR:

· All Types of Music
· Professional Sound Systems
· Light Shows · Party Props
· Big Screen Video · Karaoke

ARTIST REPRESENTATIVE:
· *American DJ Lighting*
· *PVDJ Audio*
· *Promo Only CDs & DVDs*

HAVE YOU INVITED HER YET?

Back

Front

Back

Front

Back

Brochure design
Courtesy, Dowdle Design—www.dowdledesign.com.

Postcard design
Courtesy, Dowdle Design—www.dowdledesign.com.

The *FUN* DJs

DESERT DJs

Here's what people are saying about us . . .

Our DJ was great! Our entire experience with Desert DJs from beginning to end was excellent. Recommendations will go out to all of our engaged friends.

Suzanne & Josh Goldstein
Bride & Groom

Desert DJs is the BEST!!! I love referring business to you and knowing you'll always exceed my expectations.

Carol Reid
Catering and Sales Manager
Loews Ventana Canyon Resort

We were thrilled with the great job your DJs did at Daniel's Bar Mitzvah! The DJs had unbelievable enthusiasm and lent it to the guests! With the pre-planning we did with your staff, combined with the DJs' creativity and energy, Desert DJs was the key to the entire evening! Everything went so smoothly, the DJs were so easy to work with, and I got to really enjoy the party. It was wonderful and you were perfect.

Patricia & Ladd Kleiman

We thank you for helping us arrange our reception. It was a blast! Our DJ was great. He was very flexible, dependable and upbeat. We received a lot of compliments on him and how well he kept everything going. Most of all we would like to thank Desert DJs for making our wedding reception one to remember.

Angellina & Dan Igoe
Bride & Groom

FUN GUYS...IN BOW TIES

An enthusiastic, outgoing, professional entertainer in tuxedo. That's what you'll see at every function where Desert DJs performs. "Fun Guys...In Bow Ties", means entertaining DJs dedicated to bringing fun and excitement to every occasion. Maybe you want to learn a new dance with your guests, or have a Hula-Hoop contest for the kids, or maybe Grandma wants to do the Jitterbug. Having someone to entertain who is open, willing and able to inspire these interactions is what our DJs are all about. They all attend monthly meetings where we have creative brainstorming sessions, learn new dance steps, share information on new music and stay educated in a wide variety of areas that enable each DJ to be the best in the business. To learn more about Desert DJs & our fun guys . . . call **327-2000**.

FOR THE PARTY OF A LIFETIME

Our outstanding staff will explore different ideas to make your event a unique and exciting adventure. We can also give you tips on how to make your evening run smoothly, discover options like dance step instruction, work out special announcements and fun twists. We will help to choose the perfect DJ to suit your special needs and requests. Our goal is to make sure that you are comfortable and excited from the moment you hire us through the end of your exceptionally fun event. We feature the "GTG" (great time guarantee), and a "Just ask..." policy. For your party of a lifetime, call us at **327-2000**.

Brochure from Stacy Zemon's first DJ company, which she started in the 1980s. It was co-written with her father, Bob, who was an advertising and marketing genius.

Let us turn your next party into a truly festive occasion.

Sound Performance is your 'one stop entertainment source' for the affair of a lifetime.

Select from a wide variety of affordable, top quality entertainment packages that include music spanning the decades. Our extensive record library features sounds for every taste from the 40's thru the 80's. Big Band, Disco/Dance, Oldies, Pop, Rock, Country, Ethnic, Traditional... and everything in between.

A sound for all seasons

State-of-the-art audio equipment is all we use, producing a quality of sound that's uncompromising and

unequaled. You'll not only hear the music — you'll feel the beat.

Trip the light fantastic.

A host of dazzling lighting effects are available to create a special mood for any theme — from the romantic to the fantastic. Mirror balls, strobes, traffic lights, revolving colors — even those popular bubble and fog machines.

Catch a rising star

Full color & sound videotaping of your event is also available — along with expert still photography by

highly experienced professionals. Both will be cherished for years to come as a treasured keepsake.

$50 off your first affair

We're confident that if we please you the first time, you'll call us again and again... and tell your friends about Sound Performance. But for this to happen, there has to be a *first time.* So as an added touch of friendly persuasion, we're offering you a $50 discount on your first affair. In addition, the friends and business associates you recommend to us will also be extended this same first-time courtesy.

KAREN & RICHARD'S WEDDING RECEPTION PLANNER

RECEPTION DATE: SATURDAY, MAY 19th, 2006

VENUE & VENDOR INFO.

PHOTOGRAPHER NAME: _____   GUEST ARRIVAL TIME: _____

VIDEOGRAPHER NAME: _____   BRIDAL PARTY ARRIVAL TIME: _____

WILL A MEAL BE PROVIDED FOR YOUR DJ ENTERTAINER? _____ YES _____ NO (IF YES,) THANK YOU! FOR A SIT-DOWN MEAL, KINDLY ASK YOUR VENUE CONTACT TO SEAT THE DJ AT A TABLE NEAR THE SOUND SYSTEM IF POSSIBLE.

RECEPTION HIGHLIGHTS

1. WHAT TYPE OF MUSIC WOULD YOU LIKE PLAYED FOR THE WEDDING PARTY INTRODUCTIONS?
   _____ UBEAT _____ SLOW _____ SPECIFIC REQUEST

2. WILL THERE BE A RECEVING LINE AFTER THE INTRODUCTIONS? _____ YES _____ NO

3. WILL THERE BE A BLESSING BEFORE DINNER? _____ YES _____ NO (IF YES,)
   WHO WILL GIVE IT? _____

4. WILL THERE BE A TOAST AFTER THE BLESSING? _____ YES _____ NO (IF YES,)
   WHO WILL GIVE IT? _____

5. WHEN WILL THE CAKE CUTTING TAKE PLACE?
   _____ RIGHT BEFORE DINNER _____ RIGHT AFTER DINNER

6. WHAT SONG WOULD YOU LIKE PLAYED FOR THE CAKE CUTTING CEREMONY?
   SONG TITLE & ARTIST: _____

7. **BRIDE & GROOM'S FIRST DANCE**
   SONG TITLE & ARTIST:

   _____
   - _____ RIGHT AFTER THE CAKE CUTTING
   - _____ RIGHT AFTER THE WEDDING PARTY INTRODUCTION
   - _____ RIGHT AFTER THE RECEIVING LINE
   - _____ NO BRIDE & GROOM'S FIRST DANCE

8. **WEDDING PARTY DANCE**
   SONG TITLE & ARTIST:

   _____
   - _____ HAVE THEM JOIN IN ABOUT HALF-WAY THROUGH THE BRIDE & GROOM'S FIRST DANCE
   - _____ RIGHT AFTER THE BRIDE & GROOM'S FIRST DANCE
   - _____ NO WEDDING PARTY DANCE

9. **FATHER & DAUGHTER DANCE**
   FATHER/STEPFATHER FIRST NAME:

   _____
   - _____ RIGHT AFTER THE BRIDE & GROOM FIRST DANCE
   - _____ RIGHT AFTER THE WEDDING PARTY DANCE
   - _____ NO FATHER & DAUGHTER DANCE

   SONG TITLE & ARTIST:

   _____

10. MOTHER & SON DANCE   _____ RIGHT AFTER THE FATHER & DAUGHTER DANCE

MOTHER/STEPMOTHER FIRST NAME: _____ NO MOTHER & SON DANCE

_____

SONG TITLE & ARTIST:

_____

11. BOUQUET & GARTER

_____ WE WANT BOTH THE BOUQUET AND GARTER CERERMONES (APPROX. TIME) _____

_____ WE WANT ONLY THE BOUQUET CEREMONY (APPROX. TIME) _____

_____ NO BOUQUET OR GARTER CEREMONIES (APPROX. TIME) _____

12. IF YOU HAVE CHOSEN THE BOUQUET CEREMONY, WHAT SONG WOULD YOU LIKE PLAYED?

SONG TITLE & ARTIST: _____

13. WOULD YOU LIKE THE GENTLEMAN WHO CATCHES THE GARTER TO PLACE IN ON THE LADY WHO CAUGHT THE BOUQUET? Yes_____ No_____ **(IF YES,)**

WHAT SONG WOULD YOU LIKE PLAYED?

SONG TITLE & ARTIST: _____

14. WILL THERE BE A DOLLAR DANCE?   _____ YES   _____ NO **(IF YES,)**

WHAT SONG WOULD YOU LIKE PLAYED DURING THIS EVENT?

SONG TITLE & ARTIST: _____

15. WILL THERE BE CENTERPIECES ON THE TABLES TO BE GIVEN AWAY? _____ YES _____ NO **(IF YES,)**

_____ GIVE THE CENTERPIECES AWAY TO THE PERSON AT EACH TABLE WHOSE BIRTHDAY IS CLOSEST TO OUR WEDDING DAY

_____ WE WILL HAVE A CARD/PENNY PLACED UNDER ONE PLATE AT EACH TABLE

_____ OTHER _____

16. ARE THERE ANY SPECIAL ANNOUNCEMENTS OR DEDICATIONS (E.G., ANNIVERSARIES, BIRTHDAYS, ENGAGEMENTS) YOU WOULD LIKE MADE DURING THE RECEPTION? _____ YES _____ NO

SONG TITLE & ARTIST: _____

DEDICATED TO: _____

MESSAGE: _____

SONG TITLE & ARTIST: _____

DEDICATED TO: _____

MESSAGE: _____

MUSIC & DANCING

NOTE: THE BEST WAY FOR YOUR DJ TO GET PEOPLE ONTO THE DANCE FLOOR IS TO PLAY MUSIC FOR THE DANCING MAJORITY! ON-THE-SPOT DANCEABLE REQUESTS WILL HAPPILY BE PLAYED AS SELECTION FITS AND TIME PERMITS.

1. WHEN WOULD YOU LIKE THE DANCING TO BEING? _____ AFTER DINNER _____ BEFORE DINNER

2. ON A SCALE OF 1 TO 10 (10 = MOST), HOW MUCH OF THE FOLLOWING WOULD YOU LIKE FROM YOUR DJ ENTERTAINER?

_____ Energy & Enthusiasm _____ Interaction with Guests

3. ON A SCALE OF 1 TO 10 (10 = MOST), NUMBER EACH CATEGORY FOR ITS LIKELY APPEAL TO YOUR GUESTS.

_____ SLOW SONGS _____ BIG BAND INSTRUMENTALS
_____ 70'S – 90'S DANCE CLASSICS _____ CLASSIC ROCK
_____ CURRENT DANCE HITS _____ COUNTRY
_____ OLDIES/MOTOWN _____ STANDARDS/BIG BAND
_____ ETHNIC (IRISH, ITALIAN, JEWISH, ETC.) _____
OTHER _____

4. WHICH (IF ANY) PARTICIPATION AND ETHNIC DANCES WOULD YOU LIKE PLAYED?

_____ MACARENA _____ ELECTRIC SLIDE
_____ CHICKEN DANCE _____ CHA-CHA SLIDE
_____ HOKEY POKEY _____ CONGA LINE/PARTY TRAIN
_____ ANNIVERSARY DANCE _____ HAVA NAGILA
_____ BUNNY HOP _____ TARENTELLA
_____ THE STROLL _____ DOLLAR DANCE

5. PLEASE LIST "MUST PLAY" SONGS FOR THE DANCE PORTION OF THE RECEPTION

SONG TITLE & ARTIST: _____
SONG TITLE & ARTIST: _____
SONG TITLE & ARTIST: _____
SONG TITLE & ARTIST: _____
SONG TITLE & ARTIST: _____
SONG TITLE & ARTIST: _____
SONG TITLE & ARTIST: _____
SONG TITLE & ARTIST: _____
SONG TITLE & ARTIST: _____
SONG TITLE & ARTIST: _____

6. PLEASE LIST "MUST NOT PLAY" SONGS EVEN IF REQUESTED BY A GUEST

SONG TITLE & ARTIST: _____
SONG TITLE & ARTIST: _____
SONG TITLE & ARTIST: _____
SONG TITLE & ARTIST: _____
SONG TITLE & ARTIST: _____

ENTERTAINMENT AGREEMENT

DATE MAILED: January 8, 2006

THE PARTIES: This Agreement is for Entertainment Services for the Event described below, between the undersigned purchaser of entertainment (Client) and ABC Disc Jockey Service (Company) located at 123 Rocking Lane, Boston, MA, 84745.

SERVICES: Company agrees to provide Client with a professional DJ Entertainer/Emcee (DJ), commercial sound system, music library, and deluxe light show.

CLIENT NAME: Karen Hoffman (Fiancé is Richard Bartham)

ADDRESS: 456 Marryme Drive, Boston, MA 83652   **E-MAIL:** kh@internet.com

HOME PH. (617) 000-0000   **WORK PH.:** (617) 000-0000   **CELL PH.:** (617) 000-0000

EVENT TYPE: Wedding Reception   **DATE:** Saturday, May, 19th, 2006

FACILITY: Phoenix Country Club, Presidential Room (2nd Fl.)   **PHONE NO.:** (617) 000-0000

ADDRESS: 789 Foragoodtime Street, Newton, MA 83847   **CONTACT:** Betty Muller

APPROX. NO. OF GUESTS: 150   **AGE RANGE:** 25 - 65   **AVE. AGE:** 30

$1,095 **Total Fee**

$365   **Non-refundable deposit required by January 18, 2006**

$730   **Balance due prior to completion of Event**

OVERTIME: 50% of hourly rate per half hour. When feasible, Client requests for extended playing time during Event will be accommodated. Payment is due at time of request, and may be made with Visa, MasterCard or Discover credit cards, check or cash.

GRATUITIES: Gratuities are made at the Client's sole discretion. 10% to 15% is customary for an excellent performance.

PAYMENT: Payment may be made with Visa, MasterCard or Discover credit cards, through PayPal, or by check or cash. **PLEASE MAKE ALL CHECKS PAYABLE TO ABC DISC JOCKEY SERVICE** (There will be a $20.00 fee charged for all returned checks).

CLIENT RESPONSIBILITIES: (1) To provide Company with a completed Event Planner (provided by Company), which includes a playlist of requested songs, no less than two weeks prior to the Event; **(2)** To ensure that reasonable steps will be taken to protect DJ's equipment and music while they are located at the Event facility. In the event of circumstances deemed by DJ to represent a real or implied threat of injury or harm to DJ, equipment or recordings, DJ reserves the right to cease performance until such time as Client has the threatening situation resolved; **(3)** To provide Company with quality copies of any photographs or audio-visual recordings taken at the Event where DJ is visible, within one month following the Event. Client grants Company permission to use photographs or audio-visual recordings in its marketing materials. **(4)** Client must provide Company with a minimum of four hours notice if canceling Event due to inclement weather. Rescheduling for Events canceled due to inclement weather shall be accommodated subject to Company's and DJ's availability.

- MORE ON OTHER SIDE -

PERFORMANCE: (1) DJ shall arrive at Event facility approximately one hour prior to scheduled Event start time. **(2)** In the unlikely event that DJ's performance is delayed on the date of Event, DJ shall provide Client with performance time equal to time lacking; **(3)** In the unlikely event that DJ is ill on the date of event and cannot perform the services hereunder, then Company will make every reasonable attempt to provide Client with a substitute DJ with comparable skills, sound system, music, and any other options for which Client has contracted; **(4)** Should DJ fail to appear at Event for reasons such as (a) any act of God, war, natural disaster or transportation problem over which DJ or Company has no control, (b) closure of Event venue by any local, state, or Federal Government agency, then no further payment from Client is due.

LEGAL: (1) Event facility must meet all Federal, State and Local fire and safety regulations, and hold all appropriate liability insurances, music licenses and performance permits where applicable; **(2)** Client is responsible for any damages, injuries or delays that occur as a result of failure to comply with provision number (2) of Client Responsibilities; **(3)** This Agreement cannot be cancelled or modified except in writing by either the Client or Company; **(4)** Any controversy or dispute arises out of or relating to this Agreement shall be settled in Massachusetts, according to the rules of the American Arbitration Association, and the judgment rendered by the arbitrator may be entered in any court of competent jurisdiction. The prevailing party shall be entitled to recover from the other party, all of the prevailing party's costs and reasonable attorneys' fees incurred; **(5)** This Agreement and conduct pursuant thereto shall be governed in all respects by the laws of Massachusetts.

TO CONFIRM THIS AGREEMENT: (1) SIGN ONE ENTERTAINMENT AGREEMENT AND RETURN IT TO COMPANY. Retain a copy for your records; **(2)** <u>THE DEPOSIT MUST BE RECEIVED BY THE DUE DATE TO GUARANTEE SERVICES FOR YOUR EVENT</u>; **(3)** Company is holding a tentative reservation until that date; **(4)** Agreements received after the deposit due date are subject to DJ availability; **(5)** Client signing this Agreement agrees that she/he is lawfully authorized to enter into this Agreement on behalf of his/her organization (if applicable); **(6)** This Agreement constitutes the entire agreement and understanding between the Parties with respect to the subject matter hereof.

The Parties have executed this Agreement as of the date set forth below. This Entertainment Agreement supersedes all others for the aforementioned Event date.

CLIENT SIGNATURE: _____   **Date:** _____

COMPANY SIGNATURE: _____   **Date:** _____

John Doe, President
ABC Disc Jockey Service, Inc.

FOR IMMEDIATE RELEASE

Date
Contact: John Doe
610-000-0000
No Kill Date

BRIDE CRIES AT HER OWN WEDDING

Mary Jones had been daydreaming about this special day all of her life, and she wanted it to be perfect. The bad news came soon after the bridal party arrived at the banquet hall. Part of the group waited at the entryway, ready to be introduced by the DJ hired to entertain for the wedding reception.

Looks of pity and horror swept over the faces of the three hundred guests in attendance, as the disc jockey stumbled through the introductions. First he pronounced names incorrectly. Then the music was too loud and inappropriate. Static hissed from his speakers, and, worst of all, the bride and groom were announced while Mary was in the ladies room with the Maid of Honor, touching up her makeup. When she heard what had happened, Mary burst uncontrollably into tears. Her once-in-a-lifetime celebration? Ruined by an incompetent and inexperienced DJ.

How can you avoid such a tragedy? Be a smart entertainment shopper. Choosing the right mobile disc jockey to emcee and entertain at your wedding reception is crucial to its success. Unfortunately, entertainment is one of the more common areas that couples cut back first in their budget. This all too often proves to be a crucial mistake.

The fact is that no matter how delicious the meal is or how beautiful your decorations are, if the DJ isn't great, your guests won't dance and have fun. This is why the entertainment can make or break the event. According to www.weddingwebsite.com, the average cost of a wedding ranges from $17,000 to $31,000, and the entertainment chosen is responsible for 80% of an event's success.

"So, don't be 'penny wise and pound foolish'!" advises (your name and company here). "While it's attractive to shop by price, it isn't wise to hire a mobile DJ entertainer based on this factor alone. All DJ services are not equal."

What is the best way to choose a disc jockey? Find an experienced entertainer who is a member of a professional disc jockey association and who has an extensive music library as well as professional audio and lighting equipment. (Your last name) also recommends that you "Hire an entertainer that you have already seen and liked, or ask someone whose opinion you trust for a referral. If neither of these choices is an option, call a few mobile DJ services to obtain information and rates–but do not judge value on price alone. If one company charges more than another, they may well be worth it!"

Anyone who has ever had an affair ruined like Mary Jones due to an unprofessional or ill-prepared disc jockey can attest to this fact. With a wedding, the entertainer only has one chance to get it right.

To avoid Mary's fate there are several other very important factors to consider. These are addressed in a free booklet called "(Make up a short, catchy, copyright free title)." It is written by veteran DJ and wedding entertainer, (Your full name) who regularly contributes articles to *DJ Everything* magazine. The booklet gives readers all the knowledge they need to make an informed decision as an educated consumer. To receive your copy, call (610) 000-0000. Just say, "I want a perfect wedding reception" and you'll receive their free booklet by mail. Or drop them a note at (Your company name and address).

8/19/2004

American DJ Endorses DJ Stacy Zemon

American DJ Signs Endorsement Agreement with Celebrated Mobile DJ Entertainer/Author

LOS ANGELES - American DJ has announced that it has entered into an artist-endorsement agreement with renowned mobile DJ entertainer and author Stacy Zemon. Under the terms of the 18-month contract, Zemon will be featured in the company's ads, make personal appearances for American DJ at conventions and special events, and use and endorse ADJ lighting products in her performances.

"We are honored and delighted to have such a respected industry figure as Stacy Zemon endorsing our products," said Scott Davies, General Manager of the American DJ Group of Companies. "Stacy is an industry veteran and business expert who has contributed significantly to the evolution of the mobile DJ market. She has written the book on mobile DJ-ing – literally!"

Zemon is the author of the highly acclaimed *The Mobile DJ Handbook*, the world's best-selling disc jockey book, which draws on her 25 years of experience working as a mobile, karaoke, club and radio DJ. Now in its second edition and available in a Spanish version, the book walks the reader through the entire process of starting, developing and expanding a lucrative mobile music service. A vital reference for experienced professionals, *The Mobile DJ Handbook* has helped thousands of mobiles build their businesses and incomes by covering the nuts-and-bolts issues that often evade newcomers.

As part of the artist-endorser agreement, American DJ will be running a promotion with a national charity built around *The Mobile DJ Handbook*. Details of the promotion will be featured on the company's website www.americandj.com

In addition to being an author, Zemon has been revving up partygoers and packing dance floors as one of the industry's top entertainers since 1979. And, when she's not performing at VIP functions and private gigs, Zemon consults with mobile DJs world-wide. She is currently developing a national education and certification program for disc jockeys.

For the past 10 years, Zemon has relied on American DJ products to light up her performances (and profits). "My American DJ lightshow brings crowd-moving energy to the dance floor at any event," Zemon remarked. "It increases my profit as a mobile DJ entertainer through referrals and upsell opportunities. Best of all, my American DJ gear is reliable, affordable and user-friendly." Zemon's lighting arsenal includes several American DJ lighting effects, including the Whirl 250, the Vertigo, the Quintet and the Avenger II, along with an LTS-2 black tripod stand.

This celebrity definitely has an entrepreneurial side to her personality. She once partnered with a national radio broadcasting company and formed a business to directly affiliate mobile disc jockey services within radio stations. The business grew into a multi-system operation that was the mobile DJ entertainment division of seven stations. Zemon has shared her vast industry knowledge with her business peers by speaking on the topics of prosperity, leadership, education and marketing at the *International DJ Expo* and *Mid-America DJ Convention*.

"Stacy is a multi-talented DJ, entertainer, innovator, author, teacher and entrepreneur all rolled up into one person," Davies said. "Her creative wizardry has produced an impressive record of success, and we wish her an even more exciting and prosperous future with American DJ."

For more information, call American DJ at 800-322-6337 or visit the company's website at www.americandj.com. E-mail: info@americandj.com

1/7/2005

American DJ To Donate To Ronald McDonald House

American DJ Offers The Mobile DJ Handbook On Website To Benefit Ronald McDonald House. Proceeds From DJ-Author Stacy Zemon's Best-Selling DJ Book Will Go To Family Charity

LOS ANGELES - Yo, DJs! – Want to help a good cause and further your career at the same time? Just click on American DJ's website (www.americandj.com) and order a copy of *The Mobile DJ Handbook* (Focal Press) by DJ Stacy Zemon, the world's best-selling DJ book of all time.

You'll be getting helpful tips from one of the industry's leading pros on how to make your mobile DJ business more rewarding and lucrative. And American DJ will donate a portion of the proceeds from every book sale to Ronald McDonald House Charities of Southern California, helping provide a place where families can stay while their children are being treated for life-threatening illnesses.

"American DJ has teamed up with Stacy Zemon on a charity fundraising project that we believe is a win-win situation for all concerned," said Scott Davies, General Manager of the American DJ Group of Companies. "The Ronald McDonald House provides a tremendous service and comfort to families and children during a very stressful time.

"We also believe that *The Mobile DJ Handbook* will make a big difference to members of our industry by helping them become more successful and professional in their careers," Davies added. "Stacy Zemon has literally written the book on mobile DJ-ing."

Highly acclaimed as the ultimate reference for newcomers and experienced pros alike, *The Mobile DJ Handbook* has helped thousands of mobiles build their businesses and incomes. The book, now in its second edition, provides a step-by-step guide to all aspects of starting, developing, and expanding a mobile music service—from performance and equipment, to booking, sales, and marketing. Packed with advice from well-known industry leaders, *The Mobile DJ Handbook* draws on Zemon's 25 years of experience working as a mobile, karaoke, club and radio DJ.

In addition to being one of the industry's top entertainers, Zemon is known for her business savvy. The entrepreneuring DJ once partnered with a national broadcasting company to form a mobile DJ entertainment division at a chain of radio stations. Zemon has shared her vast industry knowledge with her peers by speaking on the topics of prosperity, leadership, education and marketing at the International DJ Expo and Mid-America DJ Convention.

Zemon's book has drawn rave reviews throughout the industry. "Stacy is a dedicated and sincere woman... She has a true desire to make a difference in the mobile DJ industry," said Ray Martinez, national board member and past vice president of the American Disc Jockey Association.

The Mobile DJ Handbook is available for sale at www.americandj.com for $24.95. All books ordered via the website will have a portion of their proceeds donated to Ronald McDonald House Charities of Southern California.

For more information, or to order a copy of the book, visit the American DJ website at www.americandj.com and click on the link to *The Mobile DJ Handbook fundraiser* under "Current Specials."

Appendix

Recommended Reading

Books:

- Anything by Dale Carnegie, Wayne Dyer, Napoleon Hill, W. Clement Stone, Brian Tracy, or Zig Ziglar
- *Arts Marketing*, by Finola Kerrigan, Peter Fraser, and Mustafa Ozbilgin
- *Creating Affluence*, by Deepak Chopra, MD
- *Creative Arts Marketing*, by Elizabeth Hill, Terry O'Sullivan, and Catherine O'Sullivan
- *Do What You Love, The Money Will Follow*, by Marsha Sinetar
- *Endless Referrals*, by Bob Burg
- *First Things First; The 7 Habits of Highly Effective People*, by Stephen Covey
- *Getting Business to Come to You*, by Paul and Sarah Edwards
- *Good to Great*, by Jim Collins
- *Greatest Salesman in the World, The*, by Og Mandino
- *Guerrilla Marketing*, by Jay Conrad Levinson
- *Guerrilla Marketing Attack*, by Jay Conrad Levinson
- *How to Sell the Invisible*, by Harry Beckwith.
- *How to Work a Room: The Secrets of Savvy Networking*, by Susan Roane
- *Magic of Thinking BIG, The*, by David Schwartz
- *Making Money as a Mobile Entertainer*, by Raymond A. Mardo III
- *Manual del DJ Móvil* (*The Mobile DJ Handbook* Spanish version), by Stacy Zemon
- *Marketing: A Complete Guide in Pictures*, by Malcolm McDonald
- *Marketing Plans*, by Malcolm McDonald
- *Mobile DJ Handbook, The*, by Stacy Zemon
- *Performance Beyond Expectation*, by Ray Martinez
- *Power of Positive Thinking, The*, by Norman Vincent Peale

- *Psychology of Winning,* by Dennis Waitely
- *Raving Fans,* by Ken Blanchard and Sheldon Bowles
- *Real Magic: Creating Miracles in Everyday Life,* by Dr. Wayne W. Dyer
- *Retail Strategy,* by Jonathan Reynolds and Richard Cuthbertson
- *Selling the Invisible,* by Harry Beckwith
- *Seven Spiritual Laws of Success, The,* by Deepak Chopra
- *Spinnin' 2000,* by Bob Linquist and Dennis Hampson
- *Success Through a Positive Mental Attitude,* by W. Stone
- *Think and Grow Rich,* by Napoleon Hill
- *Turning Music into Gold,* by Jeff Mulligan, Ryan Burger
- *Whale Done,* by Ken Blanchard
- *You Can't Teach a Kid to Ride a Bike at a Seminar,* by David Sandler

Internet

- Disc Jockey News Network—www.DJNN.com
- www.djmag.com
- www.djzone.net
- www.prodj.com

Magazines:

- *Billboard* magazine—www.billboard.com
- *Dance Music Authority* magazine—www.dmadance.com
- *DJ Times* magazine—www.djtimes.com
- *Entrepreneur* magazine—www.entrepreneur.com
- *Fast Company* magazine—www.fastcompany.com
- *Inc.* magazine—www.inc.com
- *Karaoke Singer* magazine—www.eatsleepmusic.com
- *Mobile Beat* magazine—www.mobilebeat.com
- *Nightclub & Bar* magazine—www.nightclub.com
- *Promo* magazine—www.promomagazine.com
- *Remix* magazine—www.remixmag.com
- *Right On!* magazine—www.rightonmag.com
- *Singer* magazine—www.singermagazine.com
- *Special Events* magazine—www.specialevents.com
- *Spin* magazine—www.spin.com

- *PRO SL Directory*—www.ProSLDirectory.com (The Yellow Pages for Pro Sound, Lighting, DJ and Event Production)

Consultants

Ryan Burger—rburger@prodj.com
DJ Internet Consultation and Editing

Ray Mardo—raymardo@raymardo.com
www.raymardo.com
DJ Training Consultant

J. R. Silva—funtalent@aol.com
DJ Training Consultant

Sterling Valentine—sv@djprofits.com
Marketing Consultant

Sid Vanderpool—admin@djzone.com
DJ Internet Development Consultant

Stacy Zemon—djstacyz@aol.com
www.stacyzemon.com
Sales, Marketing, and Success Skills Consultant

Index

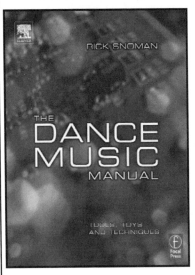